D0680501

General editor:
Brian P. FitzGerald

Science in Geography 3

Data description and presentation

Peter Davis

G. G. Aschmann

Oxford University Press 1974

Oxford University Press, Ely House, London W1

Glasgow New York Toronto Melbourne Wellington
Cape Town Ibadan Nairobi Dar es Salaam Lusaka Addis Ababa
Delhi Bombay Calcutta Madras Karachi Lahore Dacca
Kuala Lumpur Singapore Hong Kong Tokyo

© 1974 Oxford University Press

PRINTED IN GREAT BRITAIN BY OFFSET LITHOGRAPHY BY
BILLING AND SONS LTD., GUILDFORD AND LONDON

Preface

Geography in schools is at present going through a period of change, a change which represents to many a much-needed overhaul, to others an unnecessary dabble in apparently obscure and complicated statistical techniques. Society today is making demands on education which schools and colleges must recognize by taking part in certain changes, if their students are to become adults equipped to play their full part in the society of tomorrow. Much that is still taught is of questionable relevance to the student's needs.

Geography is a discipline which has been slow to carry through at school level the changes that have been taking place in universities. Because of this, ill-defined subject groups, such as environmental studies, social studies, or interdisciplinary studies, are tending to supersede it in schools. Greater co-operation between subjects is admittedly necessary, but one must be aware of the possibility of geography as a school discipline disappearing completely. This would be highly undesirable from an educational point of view, but quite deserved while geography continues to provide little intellectual rigour. The title of a subject, a mode of inquiry, or a field of knowledge is perhaps unimportant, but we do need an intellectually stronger core to what we teach and learn if the essence of geography is not to vanish from schools and colleges.

The changes that *are* beginning to be introduced are making geography more relevant to the needs of students as they become more involved in urban studies and planning, as they begin to analyse the problems of the developing countries, and as they begin to appreciate the problems of resource conservation. These changes are being accepted and becoming established in schools, but there is still very little said about the nature and philosophy of geography on which these changes depend. It is strongly felt that the sixth-form and college student should understand the more important arguments in this field. Only by involving the student in these issues can we justify what is being studied at Advanced Level and beyond.

The literary, descriptive approach to geography, where geography is treated as an arts subject, still has a definite role to play, but a clearer understanding of the nature of geography is achieved with the scientific approach. Such an approach requires the study of spatial patterns and overall systems of operation; it requires a greater degree of precision in measurement and description; it requires some estimate of the significance of inferences and conclusions drawn from the relationships being studied; and above all it requires an attempt to set up generalized theory from which predictions can be made.

The important tests of the success of the approach are:

(1) whether the student has a better understanding of the organization of society in a spatial (geographical) sense; and

(2) whether he has therefore developed a greater ability to make reasoned decisions based on his improved understanding.

On the first point, generalized 'models' or structures of the working of reality (which form the basis of scientific geography) aid understanding and act as pegs upon which to hang further ideas, concepts, and factual material. As far as the second point is concerned, a scientific approach to geography increases the ability to act upon evidence, and, through the development of general theory, allows decisions to be made which are based on a better understanding of reality. Thus courses of action can be better planned, and a more worthwhile contribution to society can be made.

The four books in the Science in Geography (S.I.G.) series are:

Developments in Geographical Method by Brian P. FitzGerald
Data Collection by Richard Daugherty
Data Description and Presentation by Peter Davis
Data Use and Interpretation by Patrick McCullagh.

The plan for the series came from an idea of Peter Bryan, from Cambridgeshire High School for Boys, whose advice during all the various stages of producing the books has been of great assistance.

Stonyhurst, August 1973 Brian P. FitzGerald

Contents

Chapter 1

The nature of geographical data

Geographical data can take many different forms–altitudes, rainfall and temperature statistics, river discharge measurements, slope angles, population figures, traffic flows, land values; the list, if not endless, is an extremely long one. Examples such as these can be given easily, but it is more difficult exactly to define the term 'geographical data'. Perhaps the easiest approach is to define 'data' first, and then to consider in what way data can be geographical.

In general terms, data are facts, the pieces of information which constitute the raw material of the subject to which they relate. There are, however, two important characteristics of data, geographical or otherwise, which distinguish them from other forms of factual information:

(a) Data are **precise, numerical facts**. The information which they give is quantitative rather than qualitative. As a result they lend themselves to statistical treatment, and are easily stored, not only as lists of figures on paper, but also on punched cards or tape, or in the memory banks of a computer.

(b) Data are usually **collected for a definite purpose**. They represent the result of time-consuming and often costly measurements and surveys. Frequently the justification for their collection is their use in testing a theory or hypothesis formulated beforehand.

Geographical data have no special characteristics by which they may be easily distinguished from data used in other subjects. It is not so much the facts themselves, as the way in which they are used and interpreted, that differentiates one subject from another. In general, we can say that any data are geographical provided that they can help to illustrate relationships, or solve problems, of a geographical nature. It does not matter, of course, that the same set of data may also interest the meteorologist or the hydrologist, the economist or the planner. Nevertheless, there are some forms of data which, by their character, are more likely to be of use to the geographer than anyone else.

Despite their variety, a common feature of most forms of geographical data is that they are **spatially distributed**, which means that their values vary from place to place. This is true whether the values relate to **points** (as with altitude, rainfall, slope angle, traffic flow), **lines** (traffic density), or **areas** (population density, crop yields). Some data sets, however, may consist of values which all relate to the same location, in which case variation in value occurs through time (climatic statistics, river discharge, mineral production). There are two further important distinctions which must be made between different types of data: **individual** and **grouped** data, and **continuous** and **discrete** data.

Individual data and grouped data

Individual data provide a precise and specific value for every item in the sample taken. They are very informative, but require laborious tabulation if the data set is large. The reports of the population census are a good example. With grouped data, individual values are not available, but it is known how many values fall into each of a number of different classes or groups, the limits of which are defined beforehand. An example is given in Fig. 1.1.

Fig. 1.1 *Size of farm holdings in England and Wales, 1967*
(**Source**: Agricultural Census 1967–8)

Size of holding (by group)	Number of holdings
0– 4·9 acres	58 911
5– 49·9 acres	117 803
50–499·9 acres	122 536
over 500 acres	7373

In this case it would obviously be impractical to list the exact size of each individual holding, since there are more than 300 000. The value of such data depends upon the number of classes used, but the information provided must obviously be more limited than with individual data. Nevertheless, grouped data may be adequate for many purposes, and sometimes they may even be preferred to individual data, provided that absolute accuracy and detail are not required. While individual data can be converted into grouped form if necessary, the reverse of this operation is obviously not possible.

Continuous data and discrete data

These terms relate to the nature of the values contained within a set of data. If there is no restriction on the value taken by each individual item in the set, other than the range between the highest value and the lowest, the data are said to be continuous. On the other hand, if the availability of values is limited, say to whole numbers, then the data are discrete.

The difference between these two types of data may be made clearer by using an example. Suppose that the highest point in an area has an altitude of 1000m, and the lowest point is at a height of 100m. An infinite number of possible altitudes lie between these two extremes, distributed along a continuous scale. Altitude is therefore said to be a continuous variable, and individual values can occur anywhere between the two limits. Suppose, on the other hand, that the largest settlement in an area has a population of 1000, and the smallest settlement contains 100 people. Population is a discrete variable, since possible values (in this case, whole numbers) lie at regular intervals along a non-continuous scale. Thus, while an altitude of 354·7m is quite possible, a population of that size is clearly not so. If, however, we were to calculate population densities for different parts of the area, the scale of possible values would be continuous; a density of 354·7, or 354·72, or even 354·726 persons per hectare is perfectly feasible.

In general we can say that data which consist of *measured values* (e.g. altitude, rainfall, river discharge, area) or *derived ratios* (e.g. population density, percentages of land use) are continuous in character. On the other hand, data which consist of *counted values* (e.g. population, livestock, vehicles, buildings) are discrete. In many cases, of course, practical limitations on the accuracy of measurement, and the need to simplify subsequent calculations, result in the taking of values to a limited number of decimal places. This rounding off process reduces the number of possible values that can be distinguished, without really altering the continuous nature of the variable. Nevertheless, it is possible to convert continuous data into discrete data by rounding off values to, say, the nearest whole number. It is obvious that all forms of grouped data must also be discrete, since class frequencies can only occur as whole numbers.

The differences outlined above are important ones, for they influence the use that can be made of the data, and govern the way in which it must be handled statistically. This will be shown in the following chapters, which deal with a variety of methods designed to present the data in a form more suitable for comparison and analysis. *It must be emphasized here that this book does not attempt to provide a comprehensive account of all the techniques that are available for this purpose. The methods described have been included because they are both effective and relatively simple, and are thus widely used.* They may be divided into two categories:

(a) methods which describe and summarize data statistically;
(b) methods which present data visually in the form of maps, diagrams, or graphs.

Chapter 2

Describing numerical distributions

Each separate value in a set of data has greatest importance when considered in relation to the other values, as part of a numerical **distribution.** The purpose of **description** is to summarize statistically the principal characteristics of such a distribution, thereby making it easier to interpret, and also aiding comparison with other distributions. In this way the complexity of the information is considerably reduced, whilst its usefulness is increased. The whole process may be regarded as one of controlled generalization about a given numerical distribution.

Fig. 2.1 *Parish sizes in part of Somerset*
(**Source**: 1961 Population Census)

Name of parish	Size in acres	Name of parish	Size in acres
Ash	1966	Martock	3807
Barwich	785	Montacute	1304
Brympton	1518	Mudford	2263
Chilthorne Domer	1484	North Perrott	1281
Chilton Cantelo	1163	Norton sub Hamdon	844
Chiselborough	797	Odcombe	1125
Closworth	3129	Rimpton	1010
East Chinnock	1350	South Petherton	3498
East Coker	1993	Stoke sub Hamdon	1381
Hardington Mandeville	2677	Tintinhull	2370
Haselbury Plucknett	2083	West Camel	1993
Ilchester	1550	West Chinnock	1161
Limington	1686	West Coker	1590
Long Load	1452	Yeovilton	2784
Marston Magna	1392	Yeovil Without	2021

A set of data relating to the size of parishes in part of Somerset is listed in Fig. 2.1.* Suppose that we wish to investigate the way in which parish sizes vary from one part of the country to another, and thus need to compare this distribution with others relating to different areas. As a first step, we may obtain a simple visual impression of the way in which the values are distributed by constructing a **histogram** as shown in Fig. 2.2. This involves first *grouping* the data into a number of classes, in this case seven, and counting the number of actual values of the variable, i.e. parish size, which occur within each class. The *frequency* of occurrences is represented on the histogram by the height of the appropriate bar. Thus, in effect, we are converting individual data into grouped form, and then illustrating it graphically.

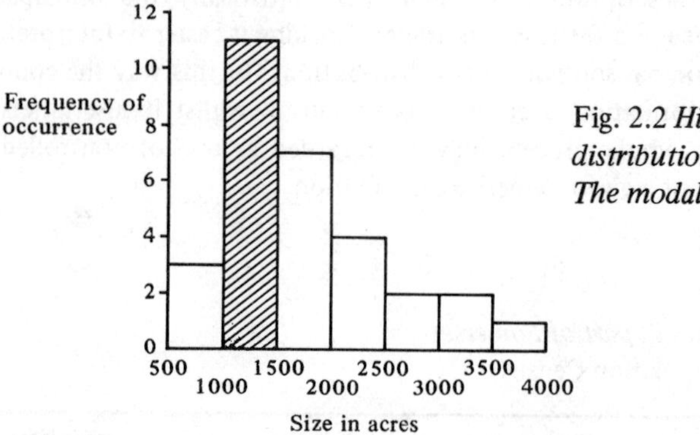

Fig. 2.2 *Histogram showing distribution of parish sizes. The modal class is shaded*

The number of classes used can vary according to the characteristics of the data involved, provided that between them they cover the whole range of values. Obviously distributions which contain large numbers of values normally require more classes than those with few. The problems of classifying data in this way will be dealt with in more detail in Chapter 5, but it is important to realise that the decision made can have a considerable effect upon the results obtained. Whatever the number of classes chosen, it is essential to the construction of the histogram that the **class interval** remains constant, or in other words that each class covers exactly the same range of values.

One way of ensuring fairly sound classification is to experiment with different class intervals, and to select the one which appears to give the

*Note that in this book most examples use data collected for official pre-metric surveys and censuses. The techniques illustrated are, of course, applicable to both metric and Imperial data.

most satisfactory result. Try this with the parish size figures, or with any other set of data. How would you expect your results to be affected if you were to increase the number of classes, or to reduce it? Are there any general principles which you might bear in mind when faced with a decision about how many classes to have? For further discussion of this point see p. 55.

A histogram provides a simple visual impression of the shape of a distribution. For instance, in the case of the histogram of parish size, there are obviously a greater number of values towards the lower end of the range, and only relatively few large parishes. Asymmetrical distributions of this sort are said to be **skewed**; in this case the histogram reveals a **positive skew**, with the **modal class**, containing the largest number of values, off centre to the left. Other distributions may exhibit a **negative skew** towards the upper end of the range, whilst many are more or less symmetrical in shape. A non-skewed distribution which is roughly symmetrical about a central modal class is referred to as a **normal distribution**. In histogram form, this would appear as in Fig. 2.3.

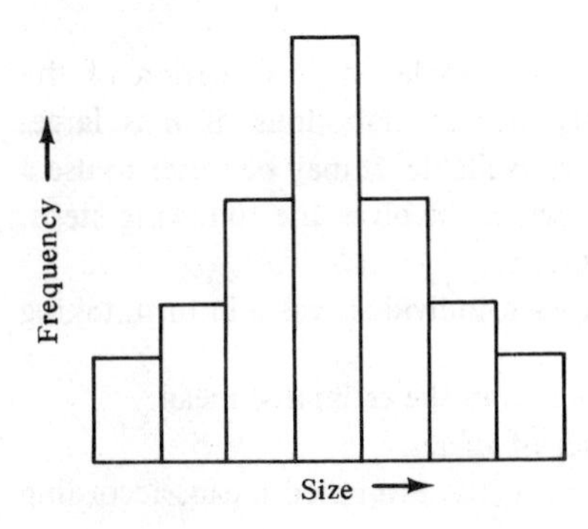

Fig. 2.3 *Histogram for a normal distribution*

Although quite useful for illustrating a distribution, the histogram does not help greatly in describing it. There are, however, a number of indices that can be calculated fairly easily, and which provide more objective descriptions of the two main properties of a distribution, namely, **central tendency** and **dispersion**.

Measures of central tendency

The first group of indices provide a number of different methods for calculating the 'average' value of a distribution. Normally we use this word rather loosely; in statistics it is important to be more specific, and to distinguish between the three different types of average that can be defined: the **arithmetic mean**, the **median**, and the **mode.**

The arithmetic mean

This index is often simply referred to as the mean, and it is the one that we most frequently imply when we 'take an average'. It is calculated by summing all the individual values of the variable, and then dividing this total by the number of values summed. Expressed mathematically,

$$\bar{x} = \frac{\Sigma x}{n}$$

where $\bar{x}$ is the mean value of the distribution, Σx (spoken as 'sigma x') is the sum of all values of x (the variable), and n is the total number of values in the distribution. Thus in the case of our parish data, the mean parish size is given by:

$$\bar{x} = \frac{53\,457}{30} = 1781{\cdot}9 \text{ acres.}$$

Although the formula itself is a very simple one, calculation of the mean by this method can obviously become laborious if n is large, especially if no calculating machines are available. It may be easier to use a common adaptation of this method, which involves the following steps.

1. Estimate a possible value for the mean.
2. Subtract the estimated value from each individual value in turn, taking careful note of the sign in each case.
3. Calculate the sum of these deviations from the estimated mean.
4. Divide this total by the total number of values.
5. Add the result to, or subtract it from, the estimated mean, according to its sign.

In practice this method involves working with smaller numbers, yet gives exactly the same result as before, regardless of the accuracy of the estimate.

Calculation of the mean in the manner described above obviously requires that the data should be available in individual form. It is possible, however, to obtain an approximate value for the mean in cases where only grouped data are available. The method used involves the setting out of the data in the form of a table, as shown in Fig. 2.4.

With grouped data, we cannot estimate an actual value for the mean, but we can predict the class in which it probably lies. In this case the second class in the table has been chosen, and this is signified by placing a zero

Fig. 2.4 *Grouped data tabulated for calculation of approximate mean. Note that the parish size data are discrete; with continuous data, both the class limits and the class mid-values would be slightly different (500–999·$\dot{9}$, with a mid-value of 750, 1000–1499·$\dot{9}$, with a mid-value of 1250, etc.)*

Class	Mid value (X_0)	Frequency (f)	Deviation from chosen class (d)	$f.d.$
501–1000	750·5	3	−1	−3
1001–1500	1250·5	11	0	0
1501–2000	1750·5	7	+1	+7
2001–2500	2250·5	4	+2	+8
2501–3000	2750·5	2	+3	+6
3001–3500	3250·5	2	+4	+8
3501–4000	3750·5	1	+5	+5
		$\Sigma f = 30$		$\Sigma fd = 31$

against this class in the column headed *d*. Deviations from the chosen class are recorded for each of the other classes in turn (note the use of the sign in Fig. 2.4). The next step is to calculate for each class the product of *f* and *d*, entering the results in the right-hand column. This column is then totalled, to give Σfd. An approximate value for the mean can then be calculated from the formula:

$$\bar{x} \simeq x_0 + c.\frac{\Sigma fd}{n}$$

where x_0 is the mid-value of the chosen class and c is the class interval. $\simeq$ means 'is approximately equal to'. Using the figures given, the approximate mean size of the parishes can be worked out as:

$$\bar{x} \simeq 1250{\cdot}5 + 500\,.\,\frac{31}{30} = 1767{\cdot}2 \text{ acres.}$$

The principle behind this method is very similar to that used when estimating a mean for individual data. The result is only approximate, however, since we are in effect assuming that the mean of all the values within a particular class is equal to the mid-value of that class, which is rarely the case. The result obtained is the same, irrespective of which class is chosen initially. This method may also be useful with large sets of individual data, where an approximate result is acceptable, and where calculation of the mean in the ordinary way may prove unnecessarily laborious.

The median

This differs from the mean in that it takes less account of each individual value in the distribution. It is defined as that value above and below which an equal number of actual values are found. Thus one way of obtaining the median is to rank all the values in order, say from highest to lowest. If the number of values in the distribution is odd, then the middle value in the ranked list is taken as the median. Thus in a set of thirty-one values, the sixteenth value in the order would be taken. In the case of the parish data, however, there are thirty values altogether, and the median must be assumed to lie mid-way between the fifteenth and sixteenth, which are 1550 and 1518 respectively. Thus we have:

$$\text{median} = \frac{1550 + 1518}{2} = 1534 \text{ acres.}$$

An alternative method of finding an approximate median value, which does not require ranking, will be explained later in the chapter.

The mode

This is defined as the most frequently occurring value within a distribution, and thus can only be found if individual values are known. In practice it is unlikely that any particular value will occur more than once or twice, unless the range of possible values is fairly small, or the data set very large. With geographical data, whether continuous or discrete, the range of values is often quite large, and any recurrence of a single value may be purely the result of chance. For instance, the sizes of parishes in Somerset vary by more than 3000 acres within the area studied, but there are two parishes with exactly the same size. In this sort of situation it is more meaningful to convert the data into grouped form, and to define instead the **modal class**, which has already been mentioned (p. 7). Reference to the histogram shows that in this case the modal class lies between the limits of 1001 and 1500.

An obvious point which arises from comparison of these three different indices of central tendency is that they each give different results. The extent to which they differ does, in fact, depend upon the nature of the distribution, and it is theoretically possible for all three to give the same value. If the distribution is more or less normal, the indices will be similar, but the gap between them increases as the distribution becomes more skewed. In this case the histogram clearly indicates a skew distribution, so it is not surprising to find a considerable difference between the median and the mean, and that both of these values lie outside the modal class.

Each of the indices described offers a different interpretation of the average value of a distribution, and is best used in conjunction with the other two. However, for purposes of comparison between two sets of data, the most valuable index is probably the mean. Unlike the other two measures, it takes into account the actual value of every item in the data, and thus has a much better mathematical basis. The mean loses some of its value in this respect if the data are in grouped form, but its main disadvantage is that for a skewed distribution it does not give a very accurate impression of the location of the majority of the values (see Fig. 2.1). While this criticism is less true of the median, and does not apply at all to the mode, both of these measures are mathematically less satisfactory in that they are derived from only one or two actual values out of the whole distribution. The modal class is especially suspect as a basis for comparison, since it is both imprecise and variable according to the classification used. Confusion may also arise if a distribution has more than one modal class.

Fig. 2.5 *A comparison of dispersion in two separate distributions*

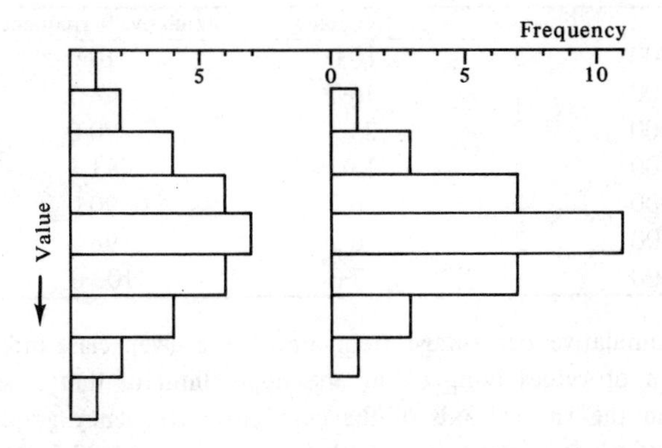

None of the indices so far mentioned is capable of *fully* describing a numerical distribution. This is illustrated in Fig. 2.5, which compares two normal distributions of the same size, with identical means, medians, and modal classes, but obviously with rather different shapes. The difference lies in the way in which individual values are spread about the mean, and it is to the description of this aspect of distribution that we must now turn.

Measures of dispersion

Dispersion is simply another word for the spread of a distribution about some average value, usually the mean. In measuring dispersion we are trying to describe how closely individual values are grouped around the mean, and once again there are a number of possible methods of approach. We shall deal with **range, mean deviation** and **standard deviation.**

Range

A very simple way of describing the dispersion of a distribution is to consider the range of values which it covers. It is much better, however, if we can get an idea of how individual values are spread within this range in relation to the average. One way of doing this is to count the number of values occuring in different parts of the range, and to illustrate the results by drawing a **cumulative frequency** graph. This can be based on the same set of grouped data as used already in constructing the histogram. In this case, however, frequencies in each class are first converted into percentages of the total number of values in the distribution, and are then **summed** to give a cumulative total (see Fig. 2.6).

Fig. 2.6 *Calculation of cumulative percentage frequency*

Class	Frequency	% frequency	Cumulative % frequency
501–1000	3	10·0	10·0
1001–1500	11	36·7	46·7
1501–2000	7	23·3	70·0
2001–2500	4	13·3	83·3
2501–3000	2	6·7	90·0
3001–3500	2	6·7	96·7
3501–4000	1	3·3	100·0

The cumulative percentage frequency for a given class indicates the proportion of values lying below the upper limit of that class. This is plotted on the vertical axis of the cumulative frequency graph, against the range of values on the horizontal axis, as shown in Fig. 2.7. By referring to the graph it is possible to divide the overall range into a number of **percentile parts**, each containing an equal proportion of the total number of values. A useful measure of dispersion is given by the size of the **inter-quartile range**, which contains the middle 50% of the values in the distribution. Approximate values for the lower and upper quartiles can be read from the graph, as can the middle value of the distribution, which is, as already explained, the median. In this case the approximate median value

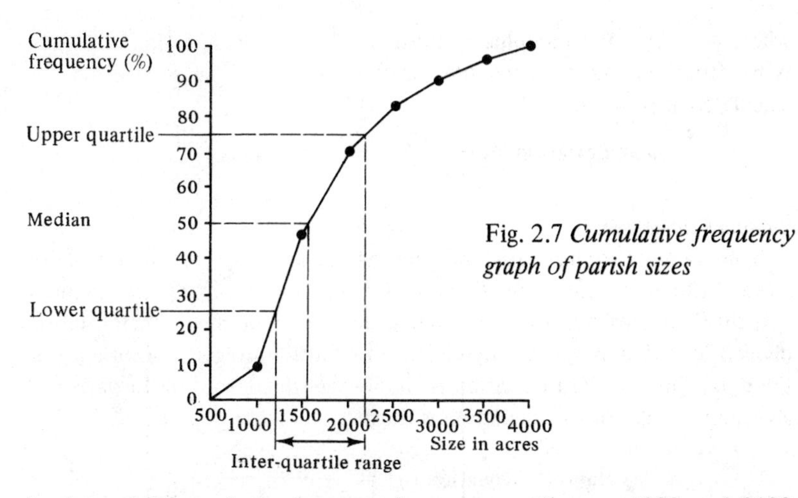

Fig. 2.7 *Cumulative frequency graph of parish sizes*

is about 1570, and the lower and upper quartiles are 1200 and 2190 respectively. Thus the size of the inter-quartile range is very approximately 1000.

If it is wished to measure the inter-quartile range more accurately, then the actual values of the quartiles must be found. We know already that fifteen values of parish size lie on either side of the median. We can now define the upper-quartile as the median of the upper fifteen values, and the lower quartile as the median of the lower fifteen. Thus the upper and lower quartiles occupy eighth and twenty-third positions respectively in the list of values ranked from highest to lowest. This gives them values of 2083 and 1281 respectively, and the inter-quartile range becomes 802. There is obviously a considerable discrepancy between the approximate and true values, but this can be reduced by plotting a smooth curve on the cumulative frequency graph instead of a series of straight lines. The chief drawback of this method of measuring dispersion, however, is that it does not consider each individual value in the distribution. It is, of course, similar in this respect to the median.

Mean deviation

This is simply a measure of the average deviation of individual values from the mean of the distribution. The mean is subtracted from each value in turn, and the result is recorded as a positive number, regardless of its actual sign (see Fig. 2.8). The results are then summed, and divided by the total number of values. This is summarized by the formula:

$$\text{Mean deviation} = \frac{\Sigma|x - \bar{x}|}{n}$$

where $|x - \bar{x}|$ is the modulus, or positive value, of the deviation of a given value from the mean. Thus the mean deviation of parish size may be calculated as follows:

$$\text{mean deviation} = \frac{18\,402{\cdot}4}{30} = 613{\cdot}4 \text{ acres.}$$

Standard deviation

This is a more sophisticated measure than mean deviation, and is slightly more difficult to calculate. Deviations from the mean are first squared (which eliminates negative deviations), and then summed. Their total is divided by the number of values in the distribution, to give the **variance** of the data. The standard deviation is simply the square root of the variance, and the formula for its calculation is thus:

$$\text{standard deviation } (\sigma) = \sqrt{\frac{\Sigma(x-\bar{x})^2}{n}}$$

The calculation of the standard deviation of the parish size distribution is illustrated in Fig. 2.8. ($\bar{x}$ is found to equal 1781·9. To find $x - \bar{x}$, 1781·9 is subtracted in turn from each value of x.) In this case we have:

$$\sigma = \sqrt{\frac{17\,499\,270{\cdot}70}{30}} = \sqrt{583\,309{\cdot}02} = 763{\cdot}7 \text{ acres.}$$

Calculation of the standard deviation by this method can be a lengthy process if the data set is large, and once again it is possible to obtain a fair approximation by using the data in grouped form. The data must first be tabulated as shown in Fig. 2.9.

The approximate value of the standard deviation is then given by the formula:

$$\sigma \simeq c\,.\sqrt{\frac{\Sigma fd^2}{n} - \left(\frac{\Sigma fd}{n}\right)^2},$$

so that, in this case:

$$\sigma \simeq 500\sqrt{\frac{101}{30} - \left(\frac{31}{30}\right)^2}$$

$$= 500\sqrt{3{\cdot}37 - 1{\cdot}07} = 500\sqrt{2{\cdot}3} = 759 \text{ acres.}$$

The standard deviation of a distribution has a variety of possible applications. For example, it is often useful to know the number of standard deviations by which an individual value in a distribution is distant from the mean. This number is known as the ***z*-score** of the value,

Fig. 2.8 *Calculation of mean deviation and standard deviation*

Size of parish (x)	$x-\bar{x}$	$\|x-\bar{x}\|$	$(x-\bar{x})^2$
1966	184·1	184·1	33 892·81
785	−996·9	996·9	993 809·61
1518	−263·9	263·9	69 643·21
1484	−297·9	297·9	88 744·41
1163	−618·9	618·9	383 037·21
797	−984·9	984·9	970 028·01
3129	1347·1	1347·1	1 814 678·41
1350	−431·9	431·9	186 537·61
1993	211·1	211·1	44 563·21
2677	895·1	895·1	801 204·01
2083	301·1	301·1	90 661·21
1550	−231·9	231·9	53 777·61
1686	− 95·9	95·9	9196·81
1452	−329·9	329·9	108 834·01
1392	−389·9	389·9	152 022·01
3807	2025·1	2025·1	4 101 030·01
1304	−477·9	477·9	228 388·41
2263	481·1	481·1	231 457·21
1281	−500·9	500·9	250 900·81
844	−937·9	937·9	879 656·41
1125	−656·9	656·9	431 517·61
1010	−771·9	771·9	595 829·61
3498	1716·1	1716·1	2 944 999·21
1381	−400·9	400·9	160 720·81
2370	588·1	588·1	345 861·61
1993	211·1	211·1	44 563·21
1161	−620·9	620·9	385 516·81
1590	−191·9	191·9	36 825·61
2784	1002·1	1002·1	1 004 204·41
2021	239·1	239·1	57 168·81

$\Sigma|x-\bar{x}| = 18\,402{\cdot}4$

$\Sigma(x-\bar{x})^2 = 17\,499\,270{\cdot}70$

and may be calculated by using the formula $z = \dfrac{|x-\bar{x}|}{\sigma}$

where the symbols are as used above. Thus the closer an individual value lies to the mean, the lower will be its z-score.

Fig. 2.9 *Data tabulated for calculating approximate standard deviation (As in Fig. 2.4, f is frequency of occurrence within class, and d is deviation from the chosen class)*

Class	f	d	d^2	fd	fd^2
501–1000	3	−1	1	−3	3
1001–1500	11	0	0	0	0
1501–2000	7	+1	1	+7	7
2001–2500	4	+2	4	+8	16
2501–3000	2	+3	9	+6	18
3001–3500	2	+4	16	+8	32
3501–4000	1	+5	25	+5	25
				$\Sigma fd = 31$	$\Sigma fd^2 = 101$

It is beyond the scope of this book to explain in detail the use of *z*-scores, but it is worth mentioning two important applications:

(a) in estimating the accuracy of the mean value of a sample (see S.I.G.s 2 and 4);

(b) in calculating the probability of individual values occurring above or below given limits, e.g. the probability of an area with a mean annual rainfall of 1000 mm receiving more than 12 000 mm of rain in a year (see S.I.G. 4).

In both these cases, use is made of the properties of the normal distribution curve, to which the distributions of many–but not all–forms of geographical data approximate. These properties are discussed in S.I.G. 4.

Measures of variability

As a final means of summarizing the characteristics of a distribution it is possible to combine measures of central tendency and dispersion in a single **variability index**. This has the advantage of allowing the dispersion of a distribution to be considered in relation to the average value about which it has been measured. Each of the three different indices used to measure variability expresses dispersion as a percentage of the average value, thus facilitating comparison between one distribution and another.

Median-based index of variability

The first index of variability has no specific title, but is based on the median value and the inter-quartile range. It is defined as:

$$\frac{\text{quartile deviation}}{\text{median}} \times 100\,\%$$

where the quartile deviation is simply half the inter-quartile range. For the parish size data it becomes

$$\frac{401}{1534} \times 100 = 26{\cdot}1\%.$$

Relative variability

This is a convenient measure which uses the values for mean deviation and the mean.

$$\text{Relative variability} = \frac{\text{mean deviation}}{\text{mean}} \times 100\,\%$$

$$= \frac{613{\cdot}4}{1781{\cdot}9} \times 100 = 34{\cdot}4\%.$$

Coefficient of variation

This index combines the values for the standard deviation and the mean, and is therefore the most sound mathematically.

$$\text{Coefficient of variation (V)} = \frac{\text{standard deviation}}{\text{mean}} \times 100\,\%$$

$$= \frac{741{\cdot}6}{1781{\cdot}9} \times 100 = 41{\cdot}6\%.$$

Each of these measures has the advantages and disadvantages of the index of central tendency and dispersion upon which it is based. Thus reliability increases through the list above, while ease of calculation decreases. The relative variability and the coefficient of variation are the two most commonly used indices, the choice between them depending upon the circumstances.

Fig. 2.10 *Summary of results for description of parish size data*

Central Tendency		*Dispersion*		*Variability*
Mode		Inter-quartile range	802	Quartile/ Median 26·1%
Modal class 1001–1500		Quartile deviation	401	Relative variability 34·4%
Median	1534	Mean deviation	613·4	Coefficient of variation (V)
Mean ($\bar{x}$): true	1781·9	Standard deviation (σ): true	763·7	41·6%
approx.	1767·2	approx.	759	

Fig. 2.10 lists the results of all the indices mentioned in this chapter, with reference to the distribution of parish sizes. Many of these indices are, of course, alternatives to one another, and in most cases it is not necessary to calculate all of them. The choice made must obviously depend upon the calculating facilities available and the degree of accuracy required of the results.

Identifying trends

With most types of geographical data, variation in quantity occurs spatially, that is from one point to another, or between areas. In some cases, however, the data may take the form of a chronological list in which variation in value occurs through time (temperature and rainfall statistics, crop yields, mineral production figures). The measures described above may be applied equally well to numerical distributions of this sort as to those in which the variation is spatial. Measures of variability are particularly valuable in this respect, especially in summarizing climatic data, but they do not indicate whether values are tending to increase, decrease, or remain fairly constant, through time. In some cases simple observation of the sequence of values may reveal apparent trends, but frequently short-term fluctuations (e.g. from one year to the next) obscure the more meaningful, underlying long-term trends. A simple solution to this problem is the use of a **running mean.**

Fig. 2.11 *Average annual yields of apples in England and Wales during the period 1948–1968. Yields are in cwts per acre.* (**Source**: Statistics published by the Ministry of Agriculture, Fisheries, & Food.

Year	Yield	Year	Yield	Year	Yield
1948	88·5	1955	76·5	1962	54·8
1949	54·0	1956	56·2	1963	86·0
1950	67·5	1957	79·7	1964	89·9
1951	63·0	1958	67·2	1965	110·6
1952	84·0	1959	97·9	1966	94·4
1953	71·7	1960	82·5	1967	71·2
1954	75·1	1961	98·7	1968	60·2

Fig. 2.11 lists the average annual yields of apples in England and Wales over a period of 21 years between 1948 and 1968. These figures are

Fig. 2.12 *Graphical representation of apple yields. Graph (a) shows actual annual yields; graph (b) is based on a 3-year running mean; and graph (c) is produced by using a 5-year running mean*

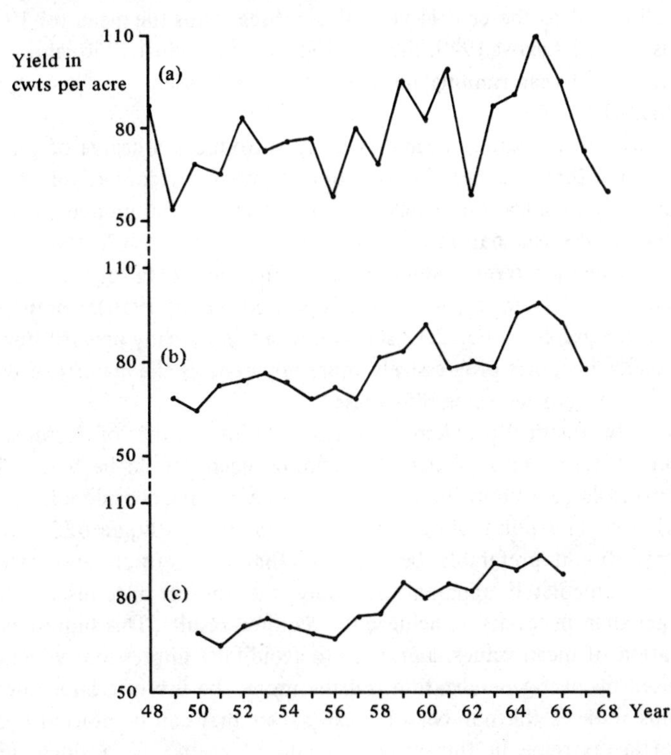

plotted as a graph in Fig. 2.12(a), which shows that short-term fluctuations in yields are very pronounced. The principal reason for such fluctuations is probably variation in climatic conditions from one year to the next. However, we may suspect that there is also a longer-term trend in yields, reflecting changes in such factors as farming practices and market requirements. Since the graph in Fig. 2.12(a) gives no clear indication of such a trend, a running mean must be used to eliminate some of the variation in the year-to-year yields recorded.

We can see from Fig. 2.11 that the yield of apples in 1948 was 88·5 cwts per acre, falling to 54·0 in 1949, and rising again to 67·5 in 1950. The mean yield for this period is obtained simply by adding the annual values and dividing by three, which gives a result of 70·0. We can calculate a mean for the period 1949–1951 in exactly the same way, arriving at a value of 61·5. Clearly mean yields can be worked out in this fashion for successive three-year periods through to 1966–1968. The results can be plotted on a graph, with the mean value for each period being allocated to the central year of the three. Thus the mean for 1948–1950 is plotted against 1949, that for 1949–1951 against 1950, and so on. This gives a **3-year running mean** trace of apple yields as illustrated in Fig. 2.12(b).

The use of a running mean obviously introduces a degree of generalization, and thus reduces the variation between successive values. The amount of generalization achieved is controlled by the number of terms included in the calculation of each mean value. Fig. 2.12(c) shows the effect of using a **5-term running mean** on the same data (in this case the first value for the period 1948–1952 is plotted against 1950). Comparison of the three graphs in Fig. 2.12 shows how a fairly steady upward trend in apple yields becomes progressively more apparent as the number of terms covered by each mean value is increased.

The only practical problem with this technique is that of deciding the number of terms upon which the running mean should be based. This will depend largely upon the nature of the data being considered, but the following general points should be kept in mind. Firstly, an odd number of terms should preferably be used, so that the central term of each period is immediately apparent. Secondly, the number of terms should be no larger than necessary to achieve the required results. This simplifies the calculation of mean values, and helps to avoid false impressions which can be caused by over-generalization. Furthermore, the use of a large number of terms reduces the number of mean values that can be plotted on the graph (the extreme in this respect would of course be a single mean value calculated for all the terms in the distribution). Thus, in the example given, there is little point in going any further than the 5-year running mean illustrated.

Although the running mean is most commonly used with data which exhibit variation through time, it is quite possible to adapt it for use in suitable circumstances within a spatial context. This is illustrated in Fig. 2.13, which presents some hypothetical land use data available for grid squares. In diagram (a) there is considerable variation in the percentage of arable land from one square to the next, but we might suspect that there is a general reduction in the amount of arable land from north-west to south-east. The existence of this trend is clearly shown in diagram (b), where 9-term running mean values are plotted. Each value is calculated for, and placed centrally within, a block of 9 grid squares. An obvious draw-

Fig. 2.13 *The use of a running mean in a spatial context. The figures are hypothetical, but diagram (a) might represent arable land as a percentage of all agricultural land in an area by grid squares. The figures in diagram (b) are 9-term running mean values, each placed centrally within the block of 9 squares to which it relates.*

(a)

68	76	73	56	64
51	82	65	61	41
71	59	75	33	67
60	69	39	56	54
47	57	28	49	61

(b)

	68·9	64·4	59·4	
	63·4	59·9	54·6	
	56·1	51·7	51·3	

back to this method is that mean values cannot be worked out for peripheral grid squares. Furthermore, it can only be used with data which are readily available in grid form. A further example of the use of a running mean with a linear, rather than areal, distribution of values is given in Chapter 8 (see p. 107).

In conclusion, it should be said that although statistical description is obviously valuable as a means of summarizing the main features of a distribution, its usefulness does not end here. Some of the indices referred to in this chapter, especially the mean and standard deviation, are of basic importance to techniques of statistical analysis, by which differences or similarities between distributions may be identified objectively. Furthermore, the properties of the normal distribution are important in establishing the statistical significance of results. Description should thus be regarded as a preliminary to analysis of some sort, rather than as an end in itself. This aspect of the interpretation of geographical data is covered more fully in S.I.G. 4.

Fig. 2.14 *Total annual precipitation in inches at two stations over a period of 21 years.* (**Source**: Records published by the Meteorological Office)

Year	Cambridge	Glasgow	Year	Cambridge	Glasgow	Year	Cambridge	Glasgow
1951	26·39	36·73	1958	29·01	37·81	1965	23·22	41·52
1952	22·09	31·83	1959	19·73	33·84	1966	23·39	43·61
1953	18·76	31·87	1960	28·52	39·22	1967	25·28	43·36
1954	24·93	51·28	1961	21·20	43·09	1968	24·60	41·28
1955	19·66	30·69	1962	19·50	42·29	1969	21·56	35·20
1956	22·04	37·07	1963	21·31	35·23	1970	22·44	41·00
1957	19·67	39·52	1964	18·52	36·68	1971	21·28	35·16

Exercises

1. Draw histograms for Cambridge and Glasgow to represent the data recorded in Fig. 2.14. What types of distribution are shown?
2. Calculate the mean and standard deviation of each of the two sets of rainfall data. Which of the two stations has experienced the greater variability of rainfall over the past 21 years?
3. Plot the data for each station on a graph. Calculate and plot 5-year running mean values on the same graphs. Are any general trends apparent for either station? Experiment with longer-term running means and compare results.
4. Construct a cumulative frequency graph for one of the stations. Read off the values of the median, and the upper and lower quartiles, and deduce the inter-quartile range. What is the variability of the rainfall assessed in this way?

Further reading for Chapter 2

Cole, J.P., and King, C.A.M., *Quantitative Geography* (Wiley, 1968), pp. 110–5 and 201–6.

Gregory, S., *Statistical Methods and the Geographer* (Longman, 1963), Chapters 2 and 3.

Monkhouse, F.J., and Wilkinson, H.R., *Maps and Diagrams* (Methuen, 1963), pp. 389–404.

Moroney, M.J., *Facts from Figures* (Penguin, 1956) Chapters 4, 5, and 6.

Theakstone, W.H., and Harrison, C., *The Analysis of Geographical Data* (Heinemann, 1970), Chapter 2.

Toyne, P., and Newby, P.T., *Techniques in Human Geography* (Macmillan, 1971), Chapter 2, especially pp. 30–45.

Chapter 3

Describing point patterns

Several of the indices used in the last chapter can be adapted for use in describing point patterns. These are, in effect, **spatial distributions**, differing from **numerical distributions** in that they are two-dimensional and cover a range of *locations* rather than *values.* This type of distribution can only contain items which are capable of representation at a point, and which are therefore discrete, rather than continuous, in space. Point patterns of possible interest to the geographer might include, amongst many others, the location of springs, wells, corries, farms, towns, factories, and shops. There are a number of ways in which patterns such as these may be described. Description should of course be regarded as a preliminary to comparison and analysis, by which the patterns may be explained.

Central tendency and dispersion

The mean centre

This is a measure of central tendency within a spatial distribution, and is calculated in a similar way to the mean of a numerical distribution. The location of a particular point in the pattern can be defined accurately by means of two co-ordinates (x, y), representing the distance of that point both horizontally and vertically from a fixed reference point. An obvious example of such a system is the National Grid, which makes it possible to define with reasonable accuracy the location of any chosen point in Britain. The mean centre of a point pattern is defined as a point which has as its two coordinates $(\bar{x}, \bar{y})$, the respective mean values of all the x and y co-ordinates in the distribution. Thus we are combining the mean values of two separate numerical distributions, scaled along different axes, to locate the mean centre of a spatial distribution.

Fig. 3.1 illustrates the location of clothes shops in the central area of Harrow. We may expect the pattern of clothes shops to be different from

Fig. 3.1 *The location of clothes shops in Harrow*

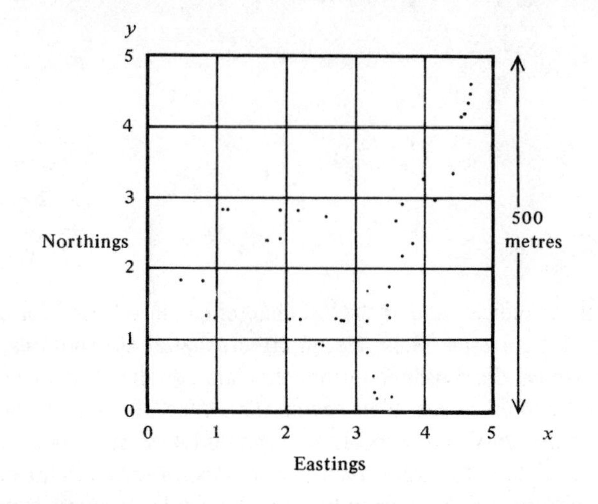

that of other establishments, such as grocers, and therefore wish to summarize this particular pattern for purposes of comparison with others. It would obviously be possible to list the x and y coordinates of the points shown quite accurately, and to calculate their separate means in the normal way. This, however, would be a fairly lengthy process, even with only forty points. It may be better, therefore, to consider the data in grouped form, using the existing grid lines as class boundaries. We thus have to count the number of points lying firstly between successive eastings, and secondly between successive northings. The results can be tabulated as shown in Fig. 3.2. This enables approximate values for the coordinates of the mean centre to be worked out using the method described in Chapter 2 (pp. 8–9). In this case the calculation is as follows:

$$\bar{x} \triangleq x + \frac{\Sigma fd}{n} \qquad \bar{y} \triangleq y + \frac{\Sigma fd}{n}$$

$$\Rightarrow \bar{x} \triangleq 3{\cdot}5 - \frac{24}{40} \qquad \Rightarrow \bar{y} \triangleq 2{\cdot}5 - \frac{16}{40}$$

$$\Rightarrow \bar{x} \triangleq 3{\cdot}5 - 0{\cdot}6 \qquad \Rightarrow \bar{y} \triangleq 2{\cdot}5 - 0{\cdot}4$$

$$\Rightarrow \bar{x} \triangleq 2{\cdot}9. \qquad \Rightarrow \bar{y} \triangleq 2{\cdot}1.$$

Thus the approximate coordinates of the mean centre are (2·9, 2·1), and this point can now be located on the map (see Fig. 3.5).

In some cases, it may be useful to attach greater importance to some points than others, perhaps by **weighting** them according to their size.

Fig. 3.2 *Tabulation of data for calculation of mean centre*

Class	Eastings			Northings		
	f	d	fd	f	d	fd
0–0·9̇	4	−3	−12	7	−2	−14
1–1·9̇	5	−2	−10	14	−1	−14
2–2·9̇	9	−1	−9	12	0	0
3–3·9̇	15	0	0	2	+1	+2
4–4·9̇	7	+1	+7	5	+2	+10
	$\Sigma f = 40$		$\Sigma fd = -24$	$\Sigma f = 40$		$\Sigma fd = -16$

Thus, in a pattern showing the locations of towns in an area, points might be weighted according to their populations, while the importance of shops or factories could take into account the number of people they employ. A point with a weighting of two, for example, would effectively count as two points, and it is the sum of these weightings between successive grid lines, rather than the actual number of points, that is used in calculating the approximate mean. The point obtained in this way is referred to as the **weighted mean centre** of the distribution, and will obviously be displaced from the true mean centre in the direction of the points with the highest weightings.

Standard distance

This index is the equivalent of standard deviation in a numerical distribution (see p. 14), and measures the degree of dispersion of points about the mean centre. It is defined thus:

$$\text{Standard distance } (S.D.) = \sqrt{\frac{\Sigma d^2}{n}},$$

where d is the distance of a given point (x, y) from the mean centre $(\bar{x}, \bar{y})$. It can be shown (by Pythagoras' theorem) that:

$$d^2 = (x - \bar{x})^2 + (y - \bar{y})^2 \quad \text{(see Fig. 3.3)}$$

and thus

$$S.D. = \sqrt{\frac{\Sigma(x - \bar{x})^2}{n} + \frac{\Sigma(y - \bar{y})^2}{n}}$$

Fig. 3.3 *Deviation of a point from the mean centre of the distribution*

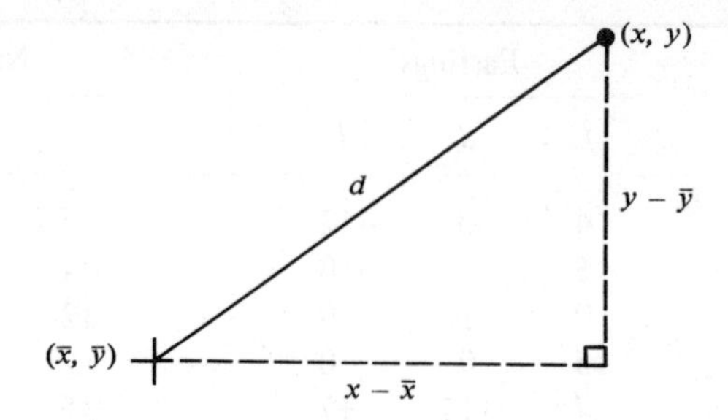

Once again it is easier to obtain an approximate result by using grouped data (see p. 2), and substituting in the formula

$$S.D. = c\sqrt{\frac{\Sigma fd_x^2}{n} - \left(\frac{\Sigma fd_x}{n}\right)^2 + \frac{\Sigma fd_y^2}{n} - \left(\frac{\Sigma fd_y}{n}\right)^2}$$

(As in formulae in Chapter 2, c is class interval).

The tabulation of data for use in this formula is illustrated in Fig. 3.4, and the final working thus becomes

$$S.D. \triangleq 1\sqrt{\frac{72}{40} - \left(\frac{24}{40}\right)^2 + \frac{64}{40} - \left(\frac{16}{40}\right)^2}$$

$$\Rightarrow \quad S.D. \triangleq \sqrt{1{\cdot}8 - 0{\cdot}36 + 1{\cdot}6 - 0{\cdot}16}$$

$$\Rightarrow \quad S.D. \triangleq \sqrt{2{\cdot}88} = 1{\cdot}7.$$

This result is in grid units, and in this case represents a distance of 170m, as each unit is 100m.

Once calculated, the standard distance is of value in that it allows the dispersion of different point patterns to be compared objectively. For in-

Fig. 3.4 *Tabulation of data for calculation of standard distance*

Class	Eastings				Northings			
	f_x	d	d^2	fd_x^2	f_y	d	d^2	fd_y^2
0–0·$\dot{9}$	4	−3	9	36	7	−2	4	28
1–1·$\dot{9}$	5	−2	4	20	14	−1	1	14
2–2·$\dot{9}$	9	−1	1	9	12	0	0	0
3–3·$\dot{9}$	15	0	0	0	2	+1	1	2
4–4·$\dot{9}$	7	+1	1	7	5	+2	4	20
				72				64

stance, the pattern of clothes shops in Harrow (or any other centre) could be compared with the pattern of butchers, or of grocers. 'Convenience' functions such as these might be expected to exhibit a greater degree of dispersion than clothes shops, which have more need of a central location. If this were so, their standard distances would obviously be larger.

In Fig. 3.5 a circle of radius equal to the standard distance has been drawn about the mean centre of the distribution of clothes shops. A feature of standard distance is that this circle will contain approximately two-thirds (in fact 68·27%) of the total number of points, provided that these are 'normally' distributed about their mean centre. In this case the fraction is slightly lower, with 25 out of 40 shops (62·5%) located within the circle marked. It follows that the pattern of clothes shops, which is clearly influenced by the layout of streets in central Harrow, is not a normal spatial distribution.

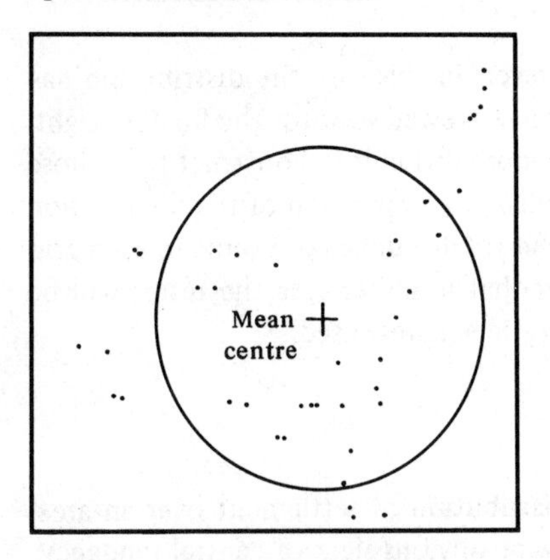

Fig. 3.5 *Mean centre and standard distance for distribution of clothes shops in Harrow*

A **normal spatial distribution** may be defined as one in which there is a symmetrical decrease in the frequency of points with increasing radial distance from a central modal area. In Chapter 2 a histogram was used to illustrate a numerical distribution of parish sizes, and to discover whether this approximated to a normal distribution (see p. 7). Histograms may be used in a similar fashion to investigate the properties of spatial distributions. One approach would be to draw separate histograms for the eastings and northings of the points in the pattern. In this case both of these would obviously exhibit skew distributions. An alternative is shown in Fig. 3.6(a), where a three-dimensional histogram is used to represent the distribution of clothes shops.

A normal spatial distribution illustrated in this way would appear as a central peak surrounded by columns becoming progressively lower towards

Fig. 3.6 *The distribution of clothes shops represented by a spatial histogram. The height of each column in diagram (a) indicates the number of shops in the grid square above which it is drawn. Diagram (b) records these totals, which can be obtained from Fig. 3.1. The mean centre is marked by a cross*

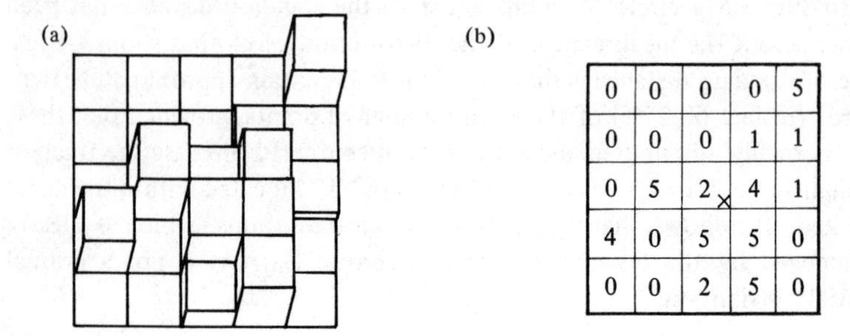

the outer edges of the pattern. However, in this case the distribution has more than one modal class, and is clearly skewed towards the bottom right. Histograms of this type are of course more difficult to construct than those used previously, and frequently an adequate impression of the distribution may be given simply by recording the total number of points in each grid square, as in Fig. 3.6(b). Remember that in either case the result will be influenced by the number of classes (grid squares) used.

The degree of clustering

Some point patterns (such as the distribution of settlement over an area, or the location of glacial erratics) show no obvious signs of central tendency. The mean centre and standard distance of such patterns, if calculated, would have no real meaning and little practical value. In describing distributions of this sort, we must therefore attempt to summarise the location of points in relation to each other, rather than to a single central point.

Examine Fig. 3.7 which shows three distributions, each containing 100 points. In case (a) the points are arranged uniformly to produce an **ordered** pattern. This is clearly different from diagram (b), where the pattern is far less regular, and shows the sort of distribution that might be produced randomly through the operation of chance. Diagram (c) shows a strongly **clustered** pattern, in which points tend to agglomerate in certain areas. Each of these patterns represents a different locational relationship between the points themselves. It is important to realize that many other patterns can exist in which these relationships are less clearly defined. In describing

(a) 100 points placed in an ordered distribution (e.g. apple trees in an orchard)

(b) 100 points in a random distribution (e.g. raindrops on a pavement)

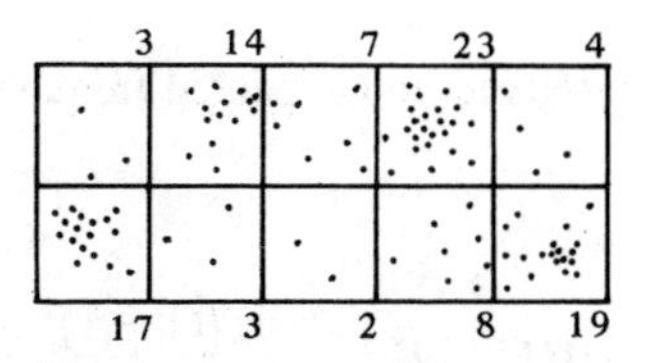

(c) 100 points in a clustered distribution (e.g. houses in a rural area of the Midlands)

Fig. 3.7 *Contrasting distributions*

them, we must try to assess objectively the degree of clustering, or the degree of uniformity, or the degree of randomness which each exhibits. Two techniques commonly used for this purpose are chi-squared analysis, and nearest neighbour analysis.

Chi-squared (χ^2) analysis

The main application of this technique is in the testing of results for statistical significance (see S.I.G. 4), but it may be adapted to the description of point patterns. If we have a perfectly ordered distribution of 100 points, and superimpose upon it a grid of 10 equal squares, we may confidently expect that each square will contain 10 points. That it does so is illustrated in Fig. 3.7(a). The basis of χ^2-analysis is the comparison of frequencies *expected* under precisely defined conditions such as these, with frequencies *observed* in the actual pattern under investigation. The procedure is as follows:

1. Count the number of points in the pattern under consideration.
2. Construct a grid of squares of equal size to cover the study area completely.
3. Calculate the **expected frequency** of points in each grid square, working on the assumption that the pattern is perfectly ordered as in Fig. 3.7(a).

$$\text{Expected frequency } (E) = \frac{\text{total no. of points in pattern}}{\text{number of grid squares}}$$

4. Count the number of points actually located within each grid square for the pattern in question, and record each total as an **observed frequency** (O). Suppose that we are investigating the pattern shown in Fig. 3.7(c). The observed frequencies in this case are the small numbers recorded at either the top right or bottom right corner of each square.
5. Calculate the value of χ^2 from the formula,

$$\chi^2 = \sum \frac{(O-E)^2}{E}$$

This is best done by tabulating the observed and expected frequencies for each square as shown below.

O	E	$O-E$	$(O-E)^2$	$\frac{(O-E)^2}{E}$
3	10	−7	49	4·9
14	10	4	16	1·6
7	10	−3	9	0·9
23	10	13	169	16·9
4	10	−6	36	3·6
17	10	7	49	4·9
3	10	−7	49	4·9
2	10	−8	64	6·4
8	10	−2	4	0·4
19	10	9	81	8·1

$$\chi^2 = \sum \frac{(O-E)^2}{E} = 52{\cdot}6$$

The value of χ^2 obtained in this way can be interpreted quite simply. If the pattern being studied is itself perfectly ordered, then clearly $\chi^2 = 0$, since expected and observed frequencies are identical. In general terms, therefore, a low total for χ^2 indicates a fairly uniform distribution of points, while a higher value suggests a greater degree of clustering. A problem which arises here is that a random distribution should also have a χ^2 value of 0, since in theory the distribution of points by chance should result in each grid square receiving the same number of points. In practice this is unlikely to occur exactly, especially if the total number of points in the pattern is fairly small. However, there will obviously be little discrimination in value at the lower end of the range between a perfectly ordered pattern and a random one. For example, the random distribution illustrated in Fig. 3.7(b) has a χ^2-value of only 4·8.

The maximum value of χ^2 is obtained when all the points in a pattern lie within one grid square. Thus in the example used here one square would contain 100 points, while the other nine contained none. By tabulating this information in the manner described above, and working through the formula, we arrive at a value of $\chi^2 = 900$. Note that this is not an absolute maximum value, but relates only to this pattern and to the grid used to cover it. An increase in either the total number of points in the pattern, or the number of grid squares used, would give a higher maximum value for χ^2.

Comparison of the value obtained in the example above ($\chi^2 = 52{\cdot}6$) with the limits mentioned is not very helpful since the value of χ^2 does not increase in direct proportion to the degree of clustering. Thus while the pattern in Fig. 3.7(c) exhibits obvious signs of clustering, its χ^2-value is relatively low. In view of this, and the problem of distinguishing between uniform and random distributions, χ^2-analysis is of little value in applying descriptive labels to individual patterns. However, **it is useful for comparing the degree of clustering present in different patterns**, its relative simplicity being an advantage in this respect. Thus patterns containing large numbers of points can be summarized and compared fairly easily without the need for a great deal of computation.

The main drawback of the method is the large variation in results that can be obtained by using different grids. Changes in the size and orientation of individual squares in the grid can have a considerable effect upon the value of χ^2, even though the number and pattern of points in the distribution remains the same. In Fig. 3.7(c), for instance, the major groupings of points are all enclosed within single grid squares; a rather lower value of χ^2 might have been expected if, instead, the grid lines had cut through these concentrations.

The size of the grid squares used determines the effectiveness of the technique in recognizing elements of clustering which may be present in a distribution. As grid squares become smaller, the discrimination of clustering improves, and the value of χ^2 therefore increases. On the other hand, the use of larger grid squares reduces the value of χ^2 and gives the impression of greater uniformity. In the extreme case, the use of a single large grid square to cover the entire study area would indicate a perfectly ordered pattern, regardless of the degree of clustering which actually existed. Thus valid comparison between distributions is only possible if a standard size of grid square is used. Special care is needed in this respect if the patterns being compared are represented on diagrams drawn to different scales.

Nearest neighbour analysis

Unlike the previous technique, this method considers the location of individual points within a pattern in relation to others. It is based on the measurement of distance between each point and its nearest neighbouring point. In a clustered distribution these distances will obviously be low, while an ordered pattern will exhibit relatively high spacings between points. To standardize results, and thus allow comparison between different point patterns, the overall density of points in the area is taken into account in the calculation of the nearest neighbour index. The procedure is as follows:

1. Locate the points in the pattern which is to be analysed on a map. Fig. 3.8 illustrates the distribution of thirteen towns in part of East Anglia. The towns, which have been classified by their functional content (see Exercise 1, Chapter 5), all act as service centres for the area surrounding them. Since each town should require roughly the same size of area to support it (assuming no great variation in population density), we may expect that the pattern of service centres should be fairly uniform.

2. Measure the distance between each town and its nearest neighbour. Frequently, only settlements of similar size or class are considered as nearest neighbours, and population may be used as an alternative to functional content in establishing those which qualify. In this example settlements of higher classes (e.g. cities) might be included as well, since these presumably also perform the functions of towns. If any settlements within the study area have nearest neighbours outside it, these can be included, provided

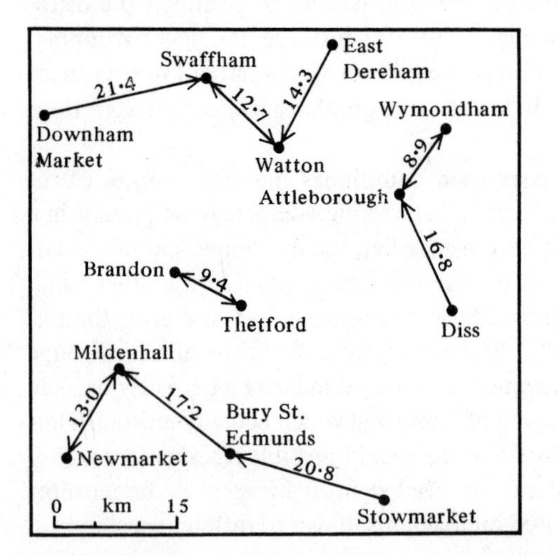

Fig. 3.8 *The nearest neighbour pattern of towns in East Anglia*

that the necessary information is available for them. In this case it is not known which centres outside the study area are towns, and so they must be ignored. Nearest neighbour relationships and distances are shown in Fig. 3.8.

3. Calculate the mean of the distances recorded in step 2, to give the observed mean distance between towns and their nearest neighbours. In this case, observed mean distance ($\bar{d}_0$) = 13·73km. This distance may of course be used for comparison with other distributions, simply to give an idea of the relative degree of spacing between points. Thus, for instance, the observed mean distance for towns in the study area could be compared with that for villages. In the same way the spacing of different types of shops in a town centre could be investigated, perhaps by considering pedestrian distances along pavements, rather than straight line distances. The effect of roads as barriers to pedestrian movement across them could be allowed for by weighting distances as necessary. In this case, of course, it is not possible to proceed any further with nearest neighbour analysis, which depends on the use of actual—rather than 'perceived'—distance.

4. Calculate the density of points in the area:

$$\text{density } (p) = \frac{\text{number of points } (n)}{\text{area } (A)}$$

$$= \frac{13}{3600} = 0{\cdot}003\,61.$$

5. Calculate the expected mean distance between towns and their nearest neighbours in a random distribution. It can be shown that:

$$\text{expected mean distance } (\bar{d}_e) = \frac{1}{2\sqrt{p}} = \frac{1}{2\sqrt{0{\cdot}003\,61}}$$

$$= 8{\cdot}\dot{3}\text{km}.$$

6. Calculate the nearest neighbour index, which compares the observed mean distance with that expected for a random distribution:

$$\text{nearest neighbour index } (R_n) = \frac{\bar{d}_0}{\bar{d}_e}$$

$$= \frac{13{\cdot}73}{8{\cdot}33}$$

$$= 1{\cdot}65.$$

This result may be reached more directly by substituting appropriate values in the formula:

$$R_n = 2\bar{d}_0 \sqrt{\frac{n}{A}}$$

A good example of the use of nearest neighbour analysis is given by King (1962). It follows from the method of calculation that, if the pattern under consideration is itself 'random', then the nearest neighbour index must have a value of 1·00. Any tendency towards clustering will cause the observed mean distance to fall below that expected, and give a value below 1·00. A value of 0 would indicate a completely clustered distribution, in which all points are coincident. In the same way any trace of uniformity will produce a value above 1·00. It can be shown mathematically that a value of 2·15 would describe a perfectly ordered distribution, in which all the points lie at the vertices of equilateral triangles. For the pattern in question a value of 1·65 is obtained. This indicates a considerable degree of uniformity, although the distribution of towns in this area clearly does not resemble very closely the ideal hexagonal pattern suggested by Christaller (1966).

In general, this technique is more rigorous than χ^2-analysis, since it deals with individual, rather than grouped, data. The nearest neighbour index is valuable in that it allows simple objective comparison between distributions without the limitations of the previous method. Thus the distribution of towns in East Anglia may be directly compared with similar patterns in other parts of the country. At the same time, the technique provides three fixed standards against which the results obtained for individual patterns may be judged.

The only real drawback of nearest neighbour analysis is that the calculation of the index is obviously a time-consuming process when the number of points in the pattern is large. In such cases it may be better to use χ^2-analysis instead. With both techniques a degree of caution is needed in the interpretation of results, especially in relation to standard values. It is generally advisable to avoid labelling distributions as 'uniform', 'random', or 'clustered', since this introduces a subjective element, and ignores the continuous nature of the scale between standard values. Instead we might say, for instance, that a particular pattern bears a closer resemblance to a uniform distribution than a random one, as in the example above. A final point to bear in mind when dealing with geographical distributions is that the term 'random' describes the appearance of a pattern, and not necessarily the factors which produced it. Thus while a settlement pattern may resemble a random distribution, it certainly does not reflect the operation of random processes.

Exercises

1. Investigate the pattern of shops in your nearest service centre of suitable size. Work out mean centres and standard distances for the distributions of several different shop types, and attempt to explain the variations you discover between them.
2. Trace the location of farms from part of a 1″ or a 1:50 000 Ordnance Survey map. Construct a grid of suitable size, and use χ^2-analysis to investigate the pattern of farms in the area chosen. Note that a fairly large number of farms should be used (say 50 or more) for the technique to work properly.

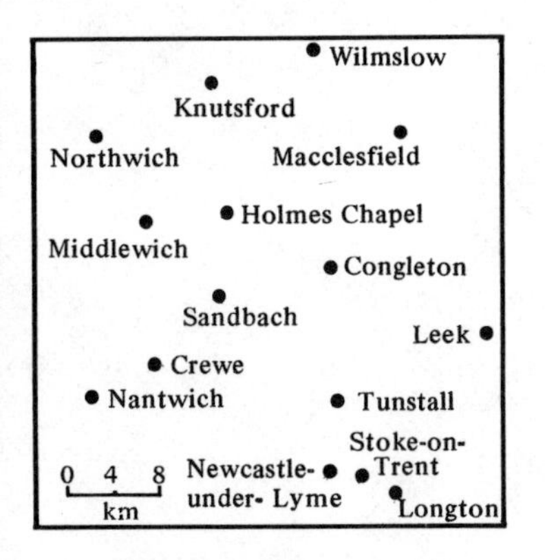

Fig. 3.9 *Settlements in the Cheshire Plain*

3. Carry out nearest neighbour analysis for the distribution of settlements shown in Fig. 3.9. What conclusions can be drawn from your result? How does it compare with that obtained for settlements in East Anglia (see p. 33)? Explain your answers.

Further reading for Chapter 3

Christaller, W., *Central Places in Southern Germany*, translated by Baskin, C.W. (Prentice-Hall, 1966).

Cole, J.P., and King, C.A.M., *Quantitative Geography* (Wiley, 1968), pp. 178–92 and 210–7.

King, L.J., 'A Quantitative Expression of the Pattern of Urban Settlements in Selected Areas of the United States', *Tijdschrift voor Economische en Sociale Geografie,* 53 (Jan. 1962), reprinted in Ambrose, P. (ed.), *Analytical Human Geography* (Longman, 1969), Section 3.

Theakstone, W.H., and Harrison, C., *The Analysis of Geographical Data* (Heinemann, 1970), Chapter 5.

Chapter 4

Describing line patterns and shape

As in the case of point distributions, it is important to give some precision to the description of transport networks, so that comparisons between one network and another may be made, and so that variations between networks may be correlated with other geographical variables.

The form of transport networks may well be related to the level of economic activity within a country and to a lesser extent to the country's connections outside its frontiers. In addition to this, the form the network takes may be closely related to the size and spacing of settlements and also to the nature of physical obstacles or barriers such as hills, estuaries, rivers and their flood-plains.

It is also only to be expected that different transport media will develop different characteristics, for example aircraft lanes are less distorted by relief than road or rail, while sea lanes have rather more dependence on physical factors such as the distribution of land and sea.

Fig. 4.1 *Rubber sheet geometry; whichever way the piece of rubber is stretched, (b) and (c), contiguity is preserved despite distortion*

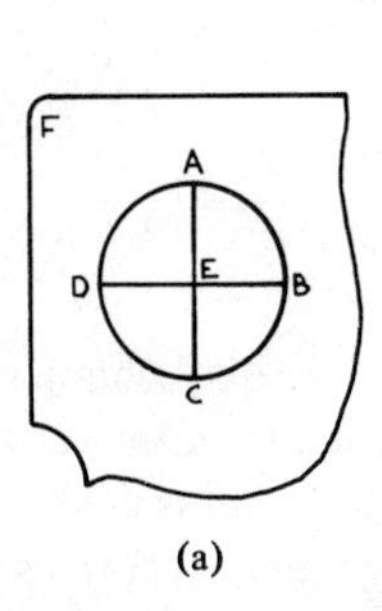

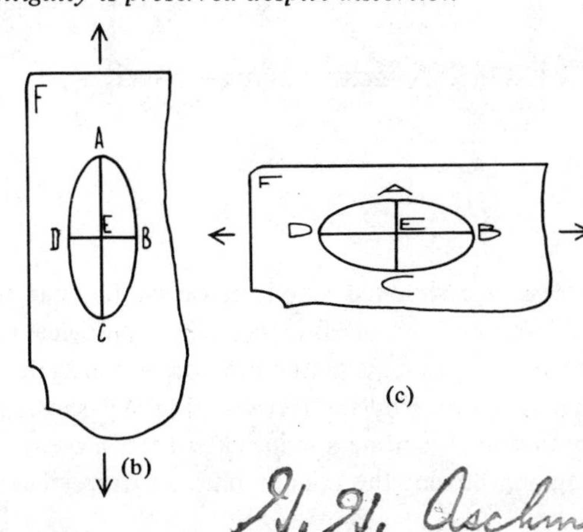

Line patterns, or networks, may be described in a variety of ways, particularly from their **topological** characteristics. These characteristics are those which do not rely upon distance and direction, but rather upon the contiguity and ordering of the lines and junctions. This part of mathematics is often referred to as 'rubber sheet' geometry. Examine Fig. 4.1, which shows a simple diagram drawn on a sheet of rubber. In whatever way the surface of the rubber is distorted, the points A, B, C, D always remain in the same order, thus preserving contiguity, while distance and direction are markedly altered. In the same way E always remains within the circuit A, B, C, D, while point F remains outside however much distortion occurs.

As contiguity is preserved throughout such apparent distortions, it is possible to study it in isolation from the other important pair of attributes of distance and direction. Thus where contiguity is of importance it is reasonable to simplify route networks so as to concentrate attention on the more important aspect of a route system as a topological network.

Fig. 4.2 *Section of the London Underground network; a topologically transformed map, preserving contiguity but not distance or direction*

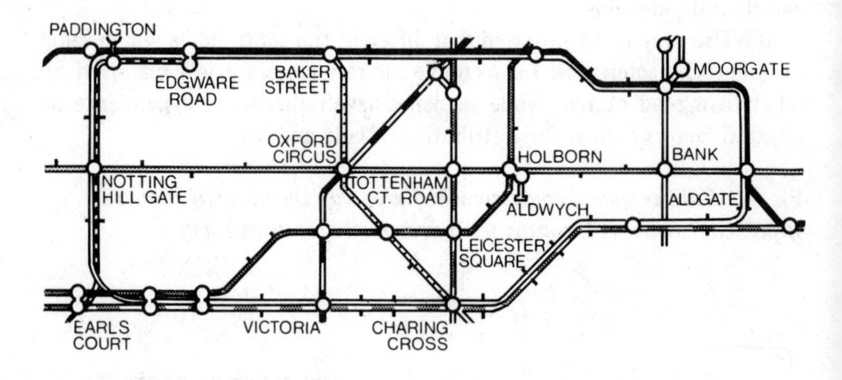

Fig. 4.2 shows a simple network familiar to travellers on London's Underground. The usefulness of the topological transformation of the actual route lies in highlighting the idea of contiguity ('Which station is next?') that is required by the traveller. Fig. 4.3 shows another topological transformation, this time a map taken from a poster of British Rail's Midland Region, showing the London line to Liverpool and Manchester.

Fig. 4.3 *Drawing taken from a British Rail poster showing a topological transformation of the map of England between London, Liverpool, and Manchester*

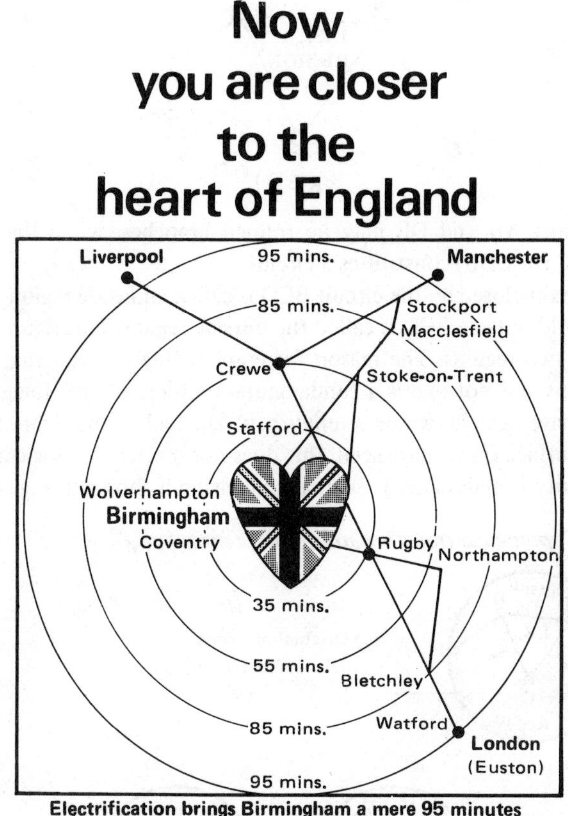

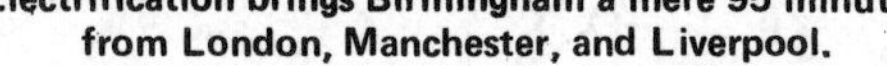

Measuring networks

Fig. 4.4 shows a simple network which could be the bus system in a small town. It consists of a series of **nodes** – A, B, C, D, and E – connected by a series of **arcs**. Elsewhere these are termed **vertices** and **edges**. A and E are **end nodes**, while C is the junction of two arcs (2-node) and B and D of three arcs (3-nodes).

Fig. 4.4 *A simple route network showing inside and outside regions*

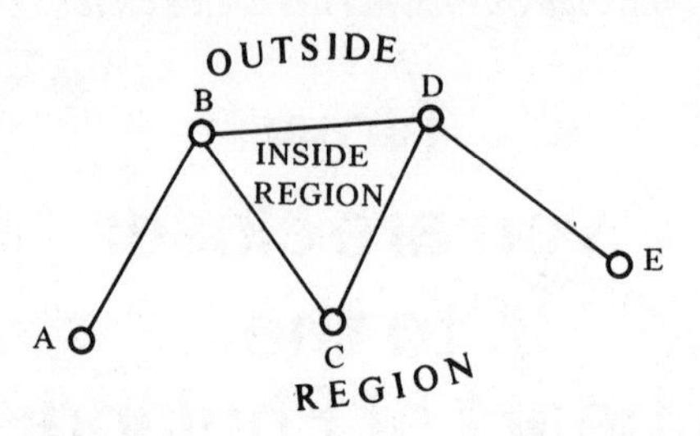

The links AB and DE may be termed **branches**, while the complete route BDC (or CDB) constitutes a **circuit**.

The area enclosed by the circuit BCD is called an **inside region** while the whole world outside BCD is called the **outside region**. This figure therefore possesses two regions. The reason for emphasizing the fact that there are two regions and not one is a fundamental topological one. Imagine such a figure to have been drawn on a tennis ball. Then allow the figure to expand over the surface (as we have seen, this does not matter topologically so long as contiguity is maintained). Such an expansion is shown in Fig. 4.5(a) and

Fig. 4.5 *Topological transformation on a tennis ball*

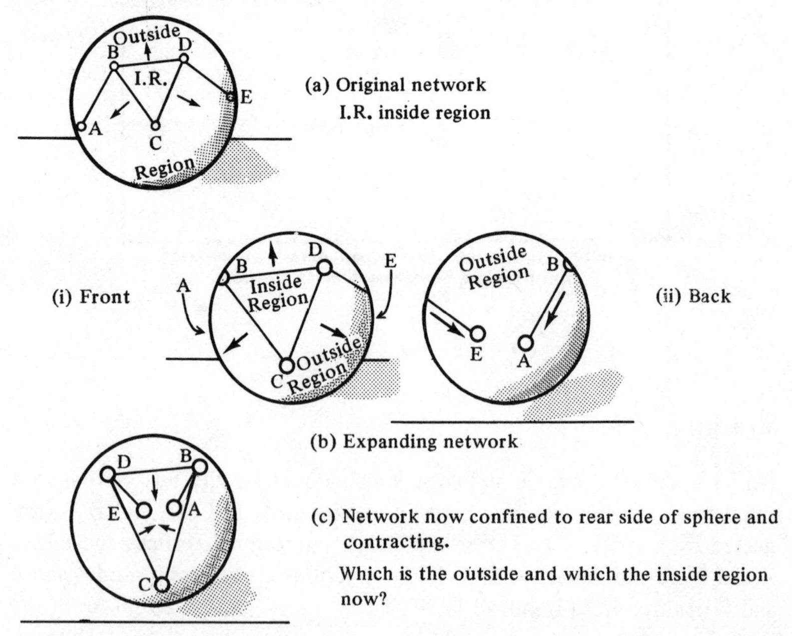

Fig. 4.6 *Network transformed by 'inversion'*

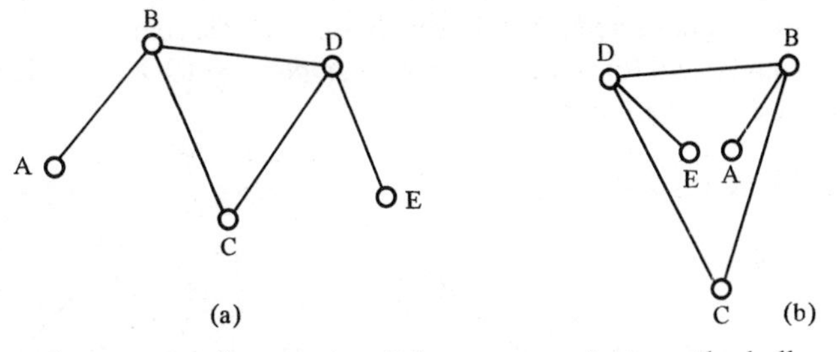

(b). At a later stage of expansion it is necessary to turn the ball round to see what is happening on the reverse side of the ball (c). In this way we arrive at the transformed figure in (c). Fig. 4.6(a) is thus topologically the equivalent of Fig. 4.6(b). What was the outside region in Fig. 4.5(a) has become the inside region in Fig. 4.5(c).

Topological transformations and their significance were discussed in S.I.G. 1, Chapter 3; here it will be sufficient to point out some of the ways in which networks may be described.

Many of these ideas were developed by K.J. Kansky, working with the railway network of Sardinia. Fig. 4.7 reproduces this network, which you may use for later exercises.

Fig. 4.7 *The rail network of Sardinia*
(After Kansky, K.J., *Structure of Transportation Networks*, Chicago Research Paper No. 84 (University of Chicago, 1963), p.8)

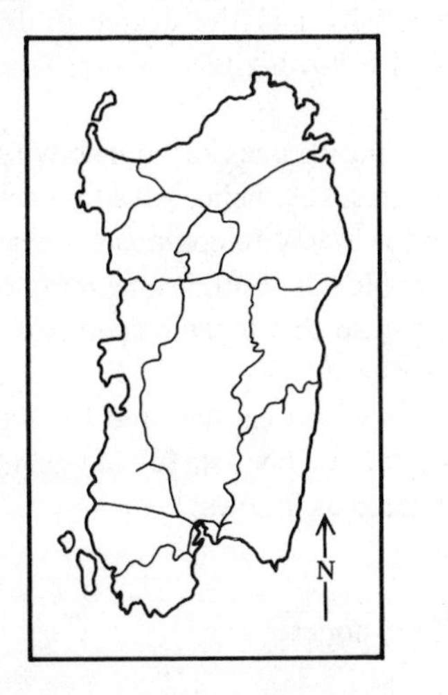

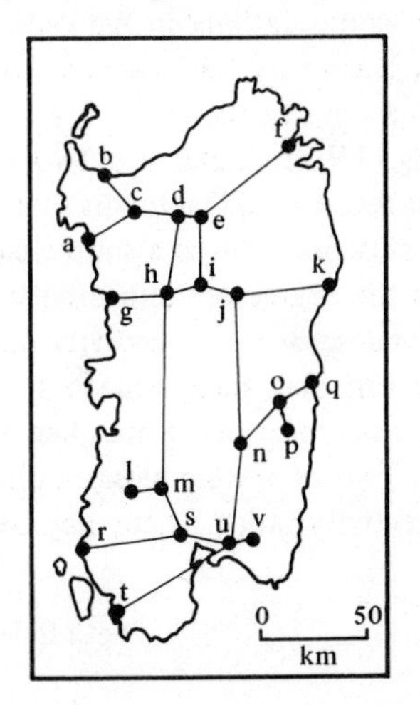

Examine Fig. 4.8. (a) shows a **null graph** (graph is the name given to any network under consideration) in which the individual nodes are not connected; (b) is a **connected graph** in which each node is connected to at least one other node; and (c) shows a **complete graph** in which each node is connected to every other node.

Fig. 4.8 *Types of graph*

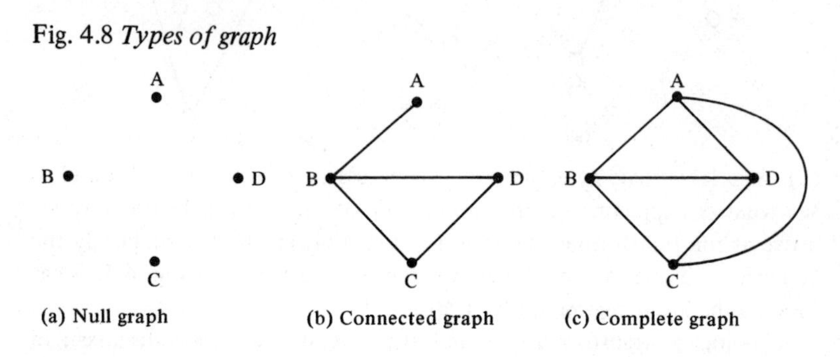

(a) Null graph (b) Connected graph (c) Complete graph

Kansky's work allows us to place numerical values on the patterns of a network from the point of view of three concepts. These concepts are: (i) connectivity (ii) centrality (iii) diameter. We shall look at them in a little more detail.

Connectivity

The degree of **connectivity** is considered to be of considerable importance in a discussion of network geography, especially as there may be some significant relationship between connectivity and the degree of development a country has reached. Exercise 3 following this chapter takes this point a stage further.

Fig. 4.9 illustrates a network in varying degrees of connectivity. The nodes A to F could be towns that are progressively being linked by a motorway network. This is a simple case, and it is easy to comment subjectively upon the degree of connectivity; the problem is to find some more precise way of describing connectivity numerically so that it can be compared with other variables, such as mean income per head.

Kansky provided a number of indices which can be used for this purpose. Two of the simplest, yet most useful, are his **beta** (β) and **gamma** (γ) connectivity ratios. These may be calculated as follows:

$$\text{beta index } (\beta) = \frac{\text{arcs}}{\text{nodes}}$$

Fig. 4.9 *Increasing connectivity (a) to (c) as a fictitious motorway network is extended between the six cities A to F*

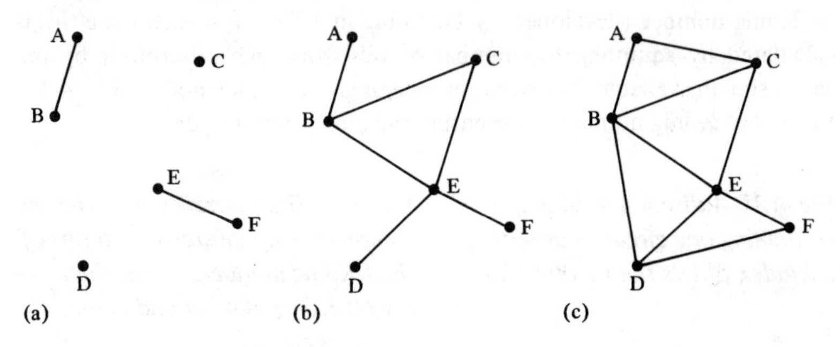

This index is designed so that any network with a beta index less than 1·00 will be composed largely of branches, while a ratio of exactly 1·00 indicates the presence of one complete circuit. A ratio of over 1·00 indicates the presence of more than one complete circuit. Note that a complete circuit could occur with a beta ratio less than 1·00 if a disconnected graph containing a circuit were being considered. (See Fig. 4.10, where the β index is 0·75).

$$\text{gamma index } (\gamma) = \frac{\text{arcs}}{3\,(\text{nodes} - 2)}$$

In this case the connectivity index always lies between 0·00, for a null graph, and 1·00 for a complete graph, a range of values being more acceptable for purposes of placing on the axis of a graph. Such graphing may be used (as in Exercise 3 at the end of this chapter) to relate the degree of connectivity of a number of countries' rail or motorway networks with their Gross National Product.

The β and γ indices for Fig. 4.9 may be calculated as follows:

	nodes	arcs	β index	γ index
(a)	6	2	$2 \div 6 = 0{\cdot}33$	$\frac{2}{3(6-2)} = \frac{2}{12} = 0{\cdot}17$
(b)	6	6	$6 \div 6 = 1{\cdot}00$	$\frac{6}{3(6-2)} = \frac{6}{12} = 0{\cdot}50$
(c)	6	9	$9 \div 6 = 1{\cdot}50$	$\frac{9}{3(6-2)} = \frac{9}{12} = 0{\cdot}75$

Centrality

The degree of **centrality** of any point on a network may be described by its **König number** (developed by D. König in 1936). For each node this is calculated by summing the number of arcs from each other node by the shortest path available. The König number is given for each node in Fig. 4.11, the lowest König number representing the most central node.

Fig. 4.10 *A disconnected graph containing one circuit, but with a β index of less than 1·00 (0·75)*

Fig. 4.11 *The centrality of nodes on a network may be stated in terms of their König numbers. E has the lowest König number and is thus the most central*

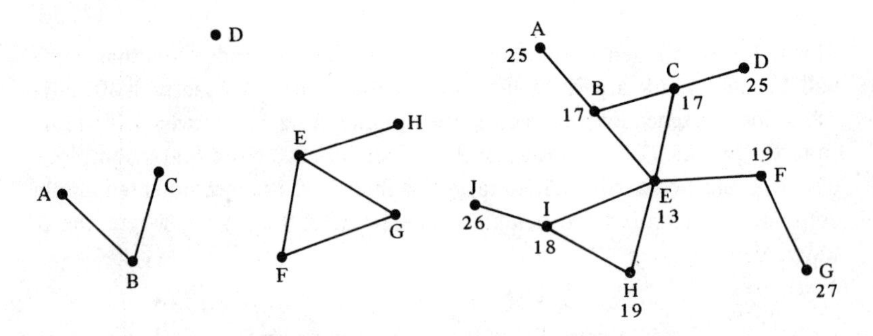

Diameter

Describing a network by means of its **diameter** involves the counting of the number of arcs in the shortest possible path between the two farthest points on the network. This is shown in Fig. 4.12. As the size of the graph increases so, in general, does the diameter (δ), although the progressive inclusion of connecting arcs causes the diameter (δ) to decrease where the number of nodes is fixed. The diameter (δ) can be related to actual distances in a network by the formula:

$$\pi = \frac{c}{d}$$ where c is the total mileage of the network
d is the mileage of its diameter (δ)

Kansky, who developed the π index as well as the β and γ indices referred to above and the η index below, called this the π index because of its analogy with the mathematically irrational number π, which is itself the ratio between a circle's circumference and its diameter.

Fig. 4.12 *Diameter (x – y) for seven simple networks. Diameter increases in general with increase in size of network [compare (a) with (g)], but the addition of connecting links may reduce the diameter [compare (e) with (d)]*

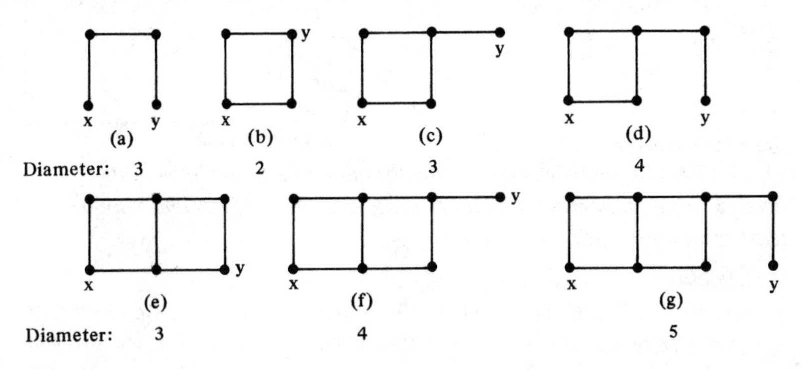

A similar index to the π ratio is Kansky's eta (η) index which also gives some idea of the spread of a network. The eta index is given by the formula:

$$\eta = \frac{c}{a}$$ where c is the total mileage of the network
a is the number of arcs.

This formula is, in effect, the average arc length of the network, and is quite sensitive to economic conditions. This can be verified by working through Exercise 3, and is elaborated by Kansky (1963, Chapter 3).

The use of matrices in network analysis

Not only may networks be described by various indices, but they may also be represented in route matrices. Examples of route matrices are fairly commonly encountered in route planning books such as the *AA Handbook of the Road.*

Examine Fig. 4.13 which shows a route system with its corresponding matrix alongside. The presence or absence of a direct link between any two nodes on the network may be represented by a '1' or a '0'. Notice how the matrix is symmetrical about the diagonal from top left to bottom right. Thus the link or arc from C to E is shown along the row C where it crosses column E, and is also repeated in row E column C. This therefore represents a link both from C to E and from E to C. This is important, as we shall see below, when considering the directional properties of a matrix.

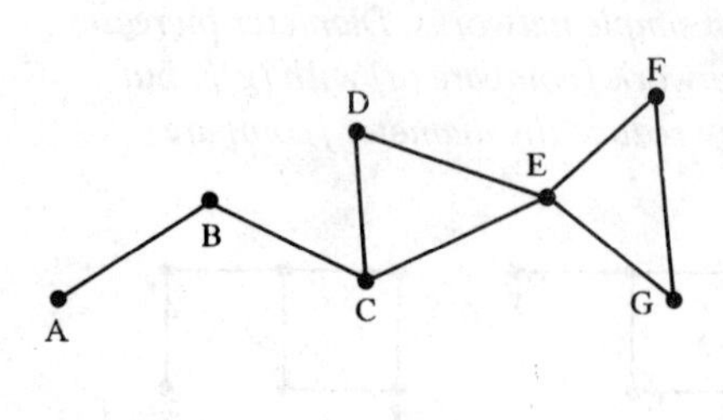

(a) Route network

From \ To	A	B	C	D	E	F	G	Total
A	0	1	0	0	0	0	0	1
B	1	0	1	0	0	0	0	2
C	0	1	0	1	1	0	0	3
D	0	0	1	0	1	0	0	2
E	0	0	1	1	0	1	1	4
F	0	0	0	0	1	0	1	2
G	0	0	0	0	1	1	0	2

(b) Route matrix

Fig. 4.13 *Route network and its corresponding matrix. The column headed 'total' indicates the number of links each node has and may be used as an indicator of accessibility*

Accessibility

The use of a route matrix, where the presence of a '1' or a '0' represents the presence or absence of a direct link between two points, can also give some idea of the **accessibility** of each node by totalling the rows (or the columns) in such a matrix. The nodes with the highest number of links as shown in the row totals may be considered to be those with the greatest degree of accessibility (note how this differs from the attribute of centrality described above). In the case of Fig. 4.13 the most accessible node would appear to be E.

The directional properties of matrices

Each row or column in a matrix can be used to represent a 'from . . . to . . .' situation. In this way row A can be used to represent 'from A to each of the places A to G'. Because of this directional property each row can be called a **row vector** and each column a **column vector**. Therefore a '1' in

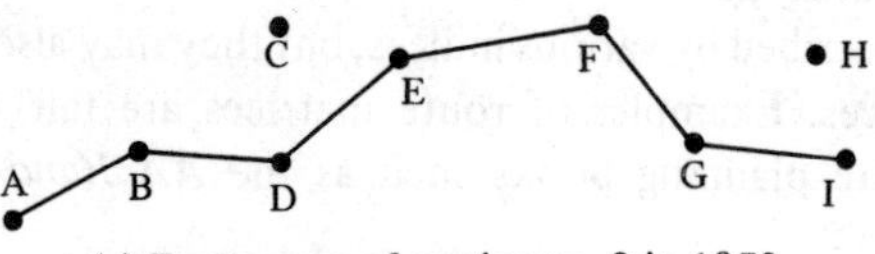

(a) Route map of service no. 9 in 1970

	A	B	C	D	E	F	G	H	I
A	0	1	0	0	0	0	0	0	0
B	1	0	0	1	0	0	0	0	0
C	0	0	0	0	0	0	0	0	0
D	0	1	0	0	1	0	0	0	0
E	0	0	0	1	0	1	0	0	0
F	0	0	0	0	1	0	1	0	0
G	0	0	0	0	0	1	0	0	1
H	0	0	0	0	0	0	0	0	0
I	0	0	0	0	0	0	1	0	0

(b) Route map of service no. 9 in 1970

Fig. 4.14 *Bus route no. 9 in a fictitious town in 1970*

row A, column C, may represent a route from A to C, but a '0' in row C, column A, would indicate the absence of a route from C back to A; in other words the route between A and C is one-way only–from A to C.

Examine Fig. 4.14 (a) where letters A to I (excluding C and H) represent bus stops on bus route number 9 in a fictitious town in 1970. Fig. 4.14(b) shows this bus route in matrix form.

In 1972 it is decided that congestion has become so great between B and E that a one-way system should be instituted via C. Also, because of the difficulty in turning the new long-wheel-base one-man operated buses in the housing estate at I, it is decided that the turn should be accomplished by routing the buses via H through I and back to G. So the bus route now becomes as shown in Fig. 4.15(a). In (b) the new bus route is shown in matrix form.

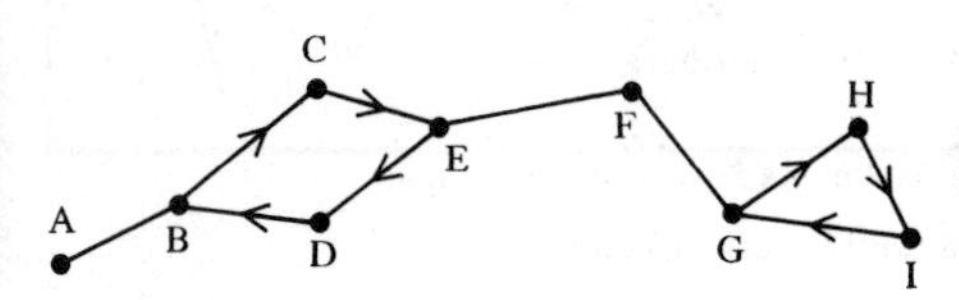

(a) Route map of service no. 9 in 1972 after inauguration of one-way systems

	A	B	C	D	E	F	G	H	I
A	0	1	0	0	0	0	0	0	0
B	0	0	1	**0**	0	0	0	0	0
C	0	0	0	0	**1**	0	0	0	0
D	0	1	0	0	**0**	0	0	0	0
E	0	0	0	1	0	1	0	0	0
F	0	0	0	0	1	0	1	0	0
G	0	0	0	0	0	1	0	**1**	**0**
H	0	0	0	0	0	0	0	0	**1**
I	0	0	0	0	0	0	**1**	0	0

(b) 1972 route matrix for service no.9

Fig. 4.15 *Bus route no. 9 after inauguration of one-way system. Changes in the matrix are marked in bold type*

Distance matrices

The matrix (which is, after all, only a convenient way to store data) may also be used for describing distances. This is shown in Fig. 4.16 where road distances between some of the more important towns in South West England are given on a route map. The matrix (b) shows the route distance, while matrices (c) and (d) are for use in the detour index exercise given below (Exercise 5), p. 53.

Detour index

We can use route distances and straight line distances to determine the efficiency of a specific route as compared with another. The following formula will produce a detour index:

Fig. 4.16 *Details of route distances and matrices for exercise (p. 53) on the effect of the Bristol Channel as a barrier*

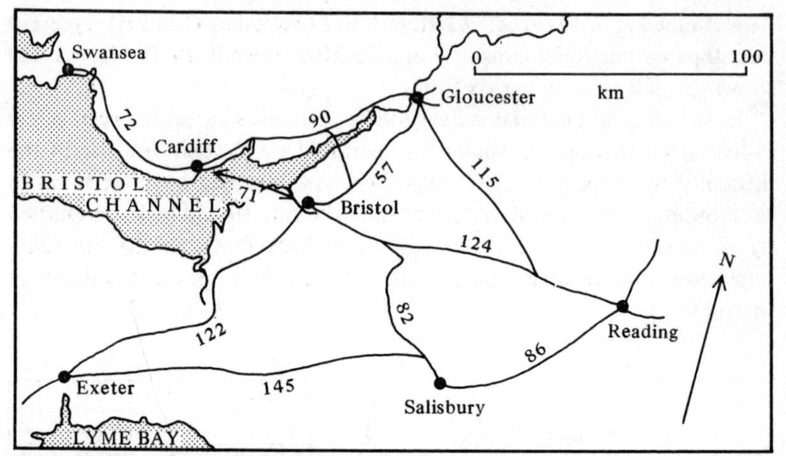

Note: Cardiff to Bristol distance is 71 km via Severn Road Bridge

(a) Map of part of South Wales and South West England

	Sw	Ca	Gl	Br	Re	Sa	Ex
Sw	0	72	162	143	267	225	265
Ca	72	0	90	71	195	153	193
Gl	162	90	0	57	115	139	179
Br	143	71	57	0	124	82	122
Re	267	195	115	124	0	86	231
Sa	225	153	139	82	86	0	145
Ex	265	193	179	122	231	145	0

Key
- Sw Swansea
- Ca Cardiff
- Gl Gloucester
- Br Bristol
- Re Reading
- Sa Salisbury
- Ex Exeter

(b) Route distance matrix

	Sw	Ca	Gl	Br	Re	Sa	Ex
Sw							
Ca							
Gl							
Br							
Re							
Sa							
Ex							

(c) Straight line distance matrix

	Sw	Ca	Gl	Br	Re	Sa	Ex
Sw							
Ca							
Gl							
Br							
Re							
Sa							
Ex							

(d) Detour index matrix

$$\text{Detour index} = \frac{\text{actual road distance}}{\text{straight line distance}} \times \frac{100}{1}$$

Such a detour index may be compared with various aspects of physical geography, such as the degree of dissection or drainage density of the area over which the routes run.

The detour index may also be used to give some comparison between routes before and after improvement has taken place, for example between the old A5 route to Birmingham and the more recent M1/M6.

Description of shape

Shape is, perhaps, the most difficult property of geographical pattern to measure. Shapes are normally subjectively described by comparison with some existing object, giving such terms as star-shaped, triangular, and dumb-bell-shaped. In other instances straightforward adjectives are used— elongated, indented, thin. None of these, however, gives a precise basis for comparison.

Shape indices

Various indices have been used to describe shape, each of which produces a ratio which may give a basis for comparison between the shape of one geographical feature and another, and also (by providing a continuous variable) gives an objective comparison with other variable quantities.

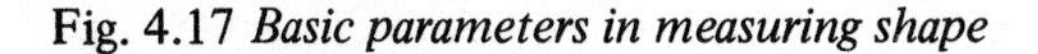

Fig. 4.17 *Basic parameters in measuring shape*

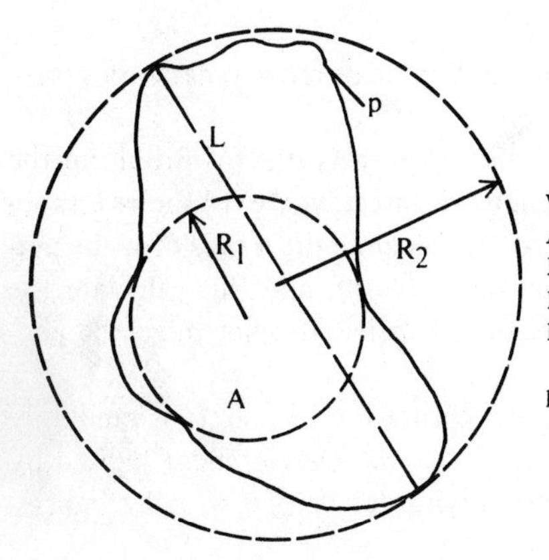

Where parameter:
- A area of shape being measured
- L length of longest axis
- R_1 radius of largest inscribing circle
- R_2 radius of smallest circumscribing circle
- p length of perimeter

Fig. 4.17 shows–for a given shape–the most easily measured **parameters.** These parameters may be combined in various ways to give ratios which will allow some precision in the comparison of different shapes.

Various indices, as described by P. Haggett (1965, pp.50–1 and 227–9), are given below:

$$S_1 = \frac{A}{0{\cdot}282\,p} \qquad S_4 = \frac{A}{\pi(0{\cdot}5\,L)^2}$$

$$S_2 = \frac{A}{0{\cdot}866\,L} \qquad S_5 = \frac{1{\cdot}27\,A}{L}$$

$$S_3 = \frac{R_1}{R_2}$$

In the case of shape index S_1 the perimeter is multiplied by 0·282, and in the case of S_2 the length is multiplied by 0·866, so that a circle will have a value of 1·0. This also applies to index S_3, where the two radii shown in Fig. 4.17 would be of the same length in the case of a circle, and thus a ratio of 1·0 would result. In the case of S_4 the index is so constructed that the area of a circle ($\pi(0{\cdot}5D)^2$) generated by a diameter of L (where the diameter, D, of a circle = the distance L) is related to the actual area of the shape under consideration. Again a circle would have a ratio of 1·0.

In each case the more linear the shape the more the ratio will tend towards zero. S_5 is treated seperately in Haggett's book, but again the ratio is constructed so that a circle will have a value of 1·0.

Exercises

1. Calculate the β and γ indices for the Sardinian railway network reproduced in Fig. 4.7.
2. Fig. 4.18(a) shows the proposed motorway network for Britain in the late 1970s, together with the main industrial areas, while (b) shows a topological transformation of this network. (i) Calculate the γ index for the network, (ii) calculate the β index for the network, and (iii) calculate the König number for each industrial area, to arrive at what might be considered as an order of centrality for each.
3. The table in Fig. 4.19 gives economic data for 18 countries randomly selected by Kansky from the *Atlas of Economic Development* (edited by N.S. Ginsburg, Chicago, 1961) together with details of β, η, and π indices. (See Kansky [1963]).

Fig. 4.18 *The main road network of England, Wales, and Southern Scotland*

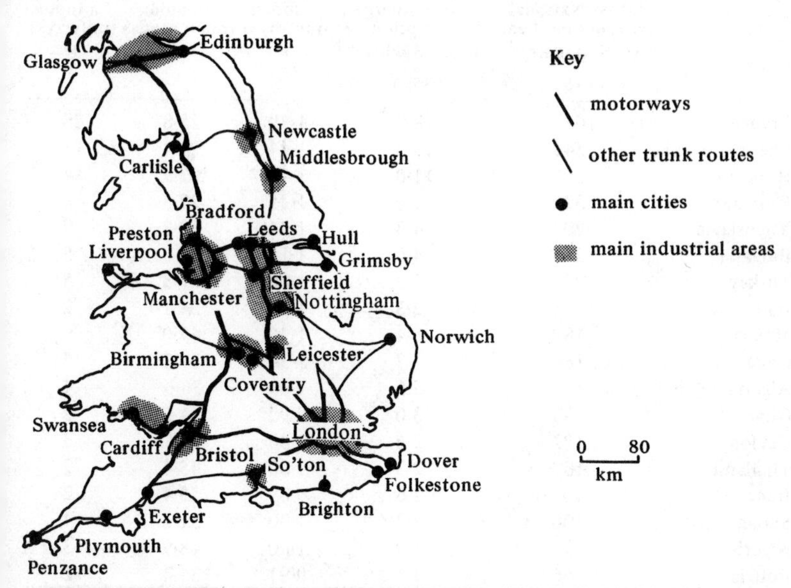

(a) Motorway and main trunk route network of Britain

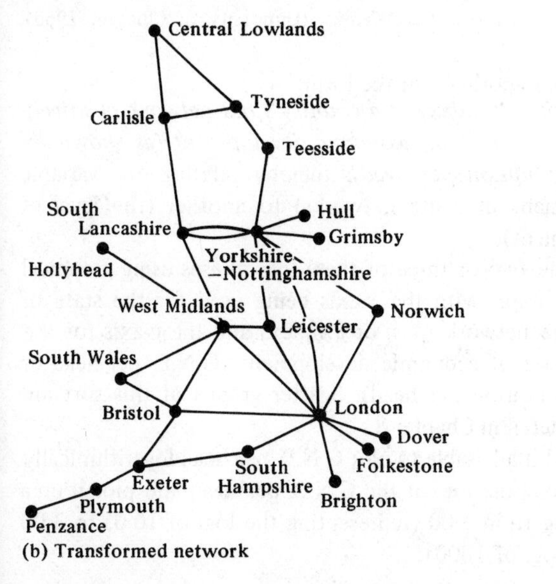

(b) Transformed network

Fig. 4.19 *Table of economic data and route network indices for 18 randomly selected countries (after Kansky)*

Country	Gross National Product per head (U.S. dollars)* (1957)	Gross Energy Consumption per head (Mw hrs)* (1956)	β index (railways) †	η index (main roads) †	π index (railways) †
France	1046	18·8	1·39	25·8	28
Czechoslovakia	543	29·0	1·44	22·4	13
Hungary	387	11·0	1·42	22·9	14
Rumania	320	5·0	1·25	27·3	8
Yugoslavia	297	4·3	1·24	32·7	9
Bulgaria	285	4·0	1·11	27·0	5
Turkey	276	3·9	1·00	27·8	5
Iraq	195	2·3	1·00	34·1	2
Mexico	187	6·4	1·25	47·0	6
Chile	180	7·7	1·34	39·2	3
Algeria	176	3·1	1·00	56·6	4
Ghana	135	3·0	0·92	44·0	2
Ceylon	122	2·6	0·88	28·5	3
Thailand	100	2·3	1·00	45·4	2
Iran	100	2·6	0·89	53·4	2
Sudan	100	2·2	1·00	51·2	4
Nigeria	70	1·9	1·00	48·9	3
Bolivia	66	3·3	0·91	35·4	2

**Atlas of Economic Development*, ed. N.S. Ginsburg (University of Chicago, 1961), pp. 18 and 80

†*Structure of Transport Networks*, K.J. Kansky (University of Chicago, 1963), pp. 42 and 44

(a) Produce a set of six hypotheses of the form: *that the connectivity (β index) of a country's rail network positively correlates with the level of its economic development (as shown by gross energy consumption per head),* thereby relating one variable (state of development of route networks) to another (the level of economic development).

(b) Proceed to examine two or three of these hypotheses using graphical methods. Draw a graph with the x-axis being used for the state of development of the network (β, η or π index) and the y-axis for the indicator of the level of economic development (G.N.P. per head or gross energy consumption per head). Scatter graphs of this sort are discussed in more detail in Chapter 8.
Note: You may find it advisable to plot G.N.P. per head logarithmically. In this case work out the log. of the G.N.P. per head and plot it on a y-axis scale ranging from 1·00 (representing the log. of 10·0) to 3·00 (representing the log. of 1000).

(c) Regression analysis and significance testing may be carried out according to the details given in S.I.G. 4 (pp.30–31).

4. Examine Fig. 4.20(a) which shows the bus stops and termini on routes 5 and 8 in the fictitious town of Market Greenhurst, and (b), which shows the route matrix for these bus routes. Make a copy of (a) and use the information contained in the matrix to work out the two bus routes. Colour the bus routes differently.

Fig. 4.20 *Details of bus routes in Market Greenhurst*

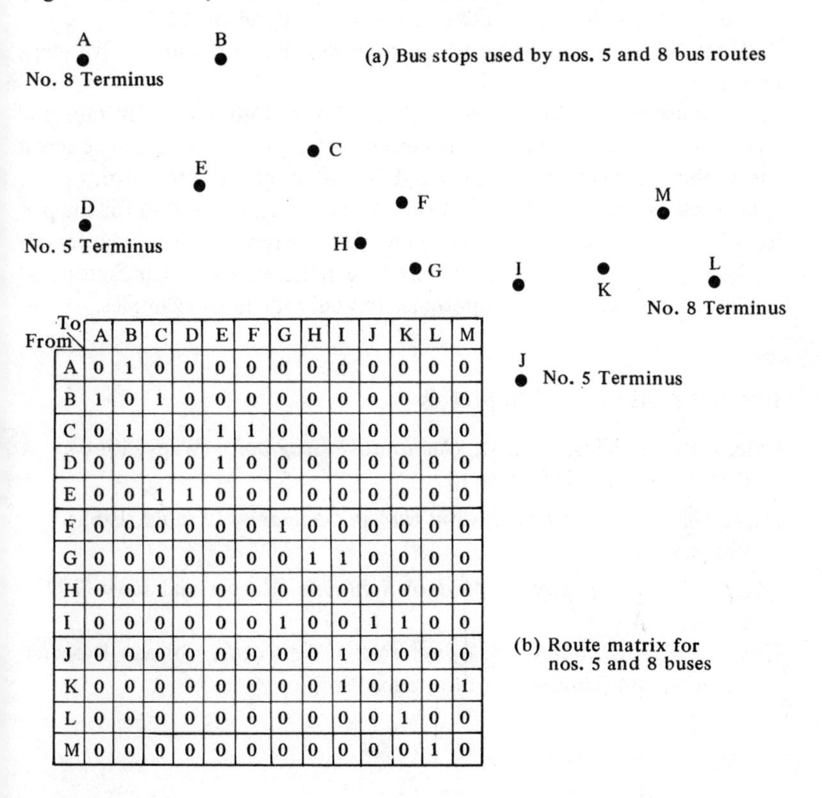

From \ To	A	B	C	D	E	F	G	H	I	J	K	L	M
A	0	1	0	0	0	0	0	0	0	0	0	0	0
B	1	0	1	0	0	0	0	0	0	0	0	0	0
C	0	1	0	0	1	1	0	0	0	0	0	0	0
D	0	0	0	0	1	0	0	0	0	0	0	0	0
E	0	0	1	1	0	0	0	0	0	0	0	0	0
F	0	0	0	0	0	0	1	0	0	0	0	0	0
G	0	0	0	0	0	0	0	1	1	0	0	0	0
H	0	0	1	0	0	0	0	0	0	0	0	0	0
I	0	0	0	0	0	0	1	0	0	1	1	0	0
J	0	0	0	0	0	0	0	0	1	0	0	0	0
K	0	0	0	0	0	0	0	0	1	0	0	0	1
L	0	0	0	0	0	0	0	0	0	0	1	0	0
M	0	0	0	0	0	0	0	0	0	0	0	1	0

5. Examine Fig. 4.16. Make a copy of each of the two empty matrices (c) and (d). Carefully measure the straight line distances between the pairs of towns and enter these in matrix (c). Now calculate the detour indices for each of the routes between pairs of towns, and enter the ratio in matrix (d).

6. An autumn gale has caused the Severn Road Bridge to be closed for a period of three days.

(a) To what extent does this alter the characteristics of the route network given in Fig. 4.16? To answer this question it is suggested that you make a new copy of the matrix (b) and then put in the new distances, and a new copy of matrix (d). Using the new route distances it is now possible to work out new detour indices for each pair of towns; remember many will remain the same. Note that the route between Cardiff and Bristol is longer with the closure of the bridge.

(b) By totalling the rows in the route distances matrices it is possible to arrive at some conclusion to the question 'which towns had the most to benefit from the construction of the Severn Road Bridge?'

7. Trace off a number of European countries and work out the five shape indices for each.

8. Make use of 1″ Ordnance Survey maps* of various parts of Britain, and trace off a number of parishes in certain well-defined areas (such as across the Chilterns, sheet 159; or in the Midlands, sheet 145 etc.; or in upland areas such as the Pennines, sheet 90). Work out shape indices for the parishes in each area, and relate these in whatever way you can to such factors as the alignment of geological outcrop, of relief, or any other factor that you may consider to be of importance in your individual examples.

Further reading for Chapter 4

Cole, J. P., and King, C. A. M., *Quantitative Geography* (Wiley, 1968). Parts 1 and 2, and Chapter 13.

Haggett, P., *Locational Analysis in Human Geography* (Arnold, 1965), Chapter 3.

Haggett, P., *Geography: A Modern Synthesis* (Harper and Row, 1972), Chapter 14.

Kansky, K.J., *Structure of Transportation Networks*, Chicago Research Paper No. 84 (University of Chicago, 1963).

*These are being replaced by the 1:50 000 maps.

Chapter 5

Classifying data

There are many circumstances in which, for the sake of simplicity, clarity, or practicality, information in grouped form is of greater value to the geographer than individual data. This is often the case when, in addition to describing the data statistically, we also wish to represent it graphically or in the form of a map. Simplicity results from the reduction of a large body of data to more easily manageable proportions. Clarity may benefit because it is often easier to identify underlying trends when the emphasis is taken off individual values. Practicality is achieved because it is impossible to differentiate between a large number of individual values when mapping, while a limited number of classes can be represented fairly easily.

The grouping of data into classes requires some care, for a poor classification can produce misleading results. Thus it is important that the method of grouping chosen should reflect as far as possible the nature of the data concerned. In this respect the main object should be to achieve:

(a) **maximum variation, and thus contrast, *between* classes;**

(b) **minimum variation, and thus maximum similarity, *within* classes.**

It is difficult to lay down hard and fast rules governing the method of classification that should be used for a particular set of data, but there are two main points which must always be taken into consideration.

1) **The number of classes.** This must obviously depend upon the total number of values in the data set. If too few classes are used, too much detail may be lost, while empty classes may result if the number chosen is too great. In general, there should be as many classes as possible without introducing empty classes, although this may be inevitable with some types of data if the full range of values is to be covered. As a rough guide the number of classes should not normally exceed five times the logarithm of the number of values in the distribution.

A set of figures relating to the input of fertilizer per hectare on sixty different farms is listed, in sequence from lowest to highest value, in Fig. 5.1. Suppose that we wish to classify this data as a first step, say, to mapping the intensity of fertilizer input in the area. Using the rule-of-thumb

Fig. 5.1 *Input of fertilizer per hectare on sixty farms*

2	5	8	13	17	22
2	6	9	13	17	23
2	6	9	13	18	25
3	6	9	14	18	26
3	6	10	14	18	26
3	7	10	14	18	27
4	7	11	14	19	28
4	7	12	16	20	31
4	8	12	16	20	32
5	8	12	17	21	36

method mentioned above, the number of classes (N) may be worked out from the formula:

$$N = 5 \times \log. 60$$
$$= 5 \times 1{\cdot}7782 = 8{\cdot}8910$$
$$= 9 \text{ (to nearest whole number)}$$

This may be regarded as the maximum number, of course, and fewer classes can be used if it appears that a better result might be obtained.

2) **The choice of class limits.** Class size obviously depends to a certain extent upon the number of classes used, and the range of values to be covered. It is important, however, that there should be some method in the definition of class limits, and the approach chosen should reflect the nature of the distribution involved, as well as the purpose for which the classification is intended.

One approach is to use the data themselves to suggest class limits, which are placed at significant values in the distribution. For instance, if values fall into a number of definable groups, it is obviously desirable to maintain the integrity of these groups in the final classification. This can be achieved by choosing **class boundaries** which coincide with recognizable breaks in the distribution. A simple method of identifying the values at which breaks occur is to plot the data along a single axis in the form of a **scattergram** (see Fig. 5.2). This shows the occurrence of individual values, and thus illustrates major groupings and obvious breaks in the distribution.

Several breaks are discernable in this example, suggesting values of 5, 11, 15, 24, and 29 for use as class boundaries. The choice is clearly subjective and some care is needed to decide where the most important breaks are located. When this method of defining class boundaries is used, it is advis-

Fig. 5.2 *Scattergram for fertilizer input data. The arrows indicate possible values for use as class boundaries*

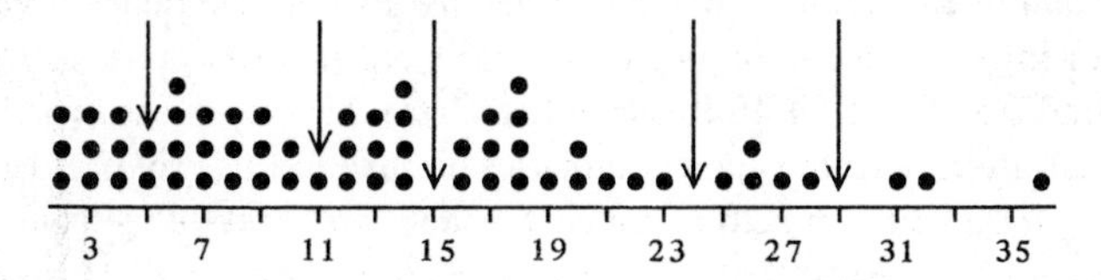

able to allow the scattergram to suggest the most appropriate number of classes; in this case the use of six classes is clearly indicated. This eliminates the problem of finding that the number of apparent breaks does not coincide with the number of class boundaries required.

Class limits must be carefully defined in order to avoid confusion over the allocation of certain values to the correct class. For instance, if in the example above the class limits are taken as 2–5, 5–11, 11–15, the values of 5 and 11 can be allocated to more than one class. This problem can be avoided by arranging the limits as 2–5, 6–11, 12–15, 16–24, 25–29, and 30–36. If the data were continuous, rather than discrete, the class limits would become 2–4·$\dot{9}$, 5–10·$\dot{9}$, 11–14·$\dot{9}$, 15–23·$\dot{9}$, 24–28·$\dot{9}$, and 29–36·$\dot{9}$, so as to include all possible values of the variable. Note that there is no overlap between successive classes in either case, and thus there is only one class to which any given value can be allocated.

An alternative approach is to plot the data in the form of a **cumulative frequency graph**, and to locate class boundaries at values where marked breaks of slope occur (see Fig. 5.3). This is harder to construct than the

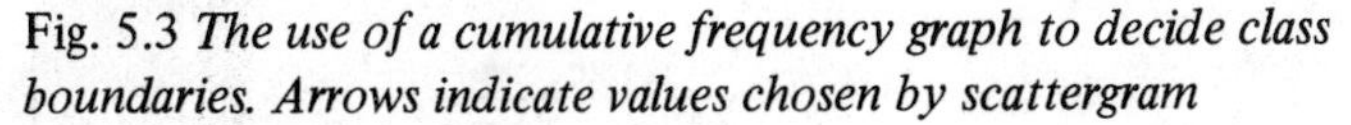

Fig. 5.3 *The use of a cumulative frequency graph to decide class boundaries. Arrows indicate values chosen by scattergram*

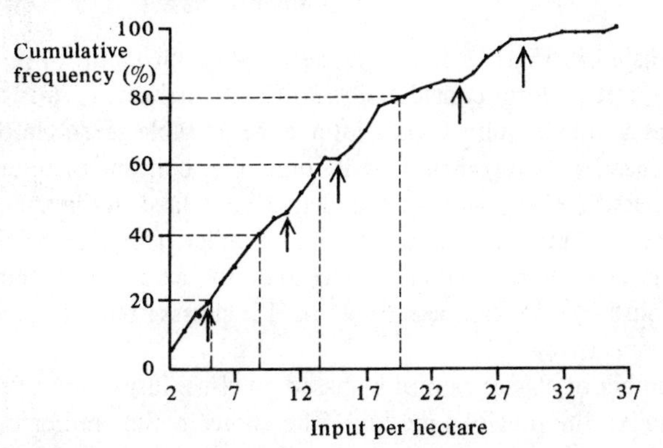

scattergram, and may not illustrate breaks in the distribution so clearly. It can, however, be used in another way to delimit classes which contain an equal number of values. This is achieved by dividing the range of the distribution into a number of percentile parts equal to the number of classes required (see Fig. 5.3). In this case five classes have been used for purposes of illustration, and the class boundaries defined in this way occur at values of 5·2, 9·0, 13·8, and 20·6. Since the data are discrete we can use class limits of 2–5, 6–9, 10–13, 14–20, and 21–36. Each of the classes defined in this way contains a fifth of the total number of values. This method is useful with data which cannot easily be split up into separate groups, since it at least ensures that all classes are roughly equal in importance.

Class intervals

A feature of the methods of classification described above is that the **class interval** can vary freely to suit the characteristics of the data. This is not the case with other, more formal methods of defining classes, which use either a fixed interval, or an interval which varies in some regular manner.

Fixed intervals

In this case the numerical extent of each class is the same, so that class boundaries are regularly spaced throughout the range of the data. The interval to be used is clearly determined by the range of values, and by the number of classes chosen. It may be calculated easily from the formula:

$$\text{class interval} = \frac{\text{range}}{\text{number of classes}}$$

The data listed in Fig. 5.1 have a range of 34 (2–36), but for purposes of classification it is better considered as 35 (2–36 inclusive). Thus the use of five classes would require a class interval of 7, while seven classes would obviously need an interval of 5. It is sometimes convenient to extend the range of classification slightly beyond that contained in the data, in order to obtain a class interval which is a round number. This is especially useful if, as in this case, discrete data are being used. By increasing the range to 36 (2–37 inclusive), we can use 6, 9, or 12 classes, with intervals of 6, 4, and 3 respectively.

The number of classes cannot be based on visual inspection of the distribution, unlike the previous method. The choice is thus rather more arbi-

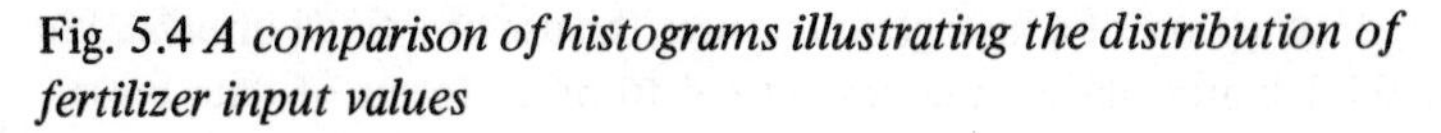

Fig. 5.4 *A comparison of histograms illustrating the distribution of fertilizer input values*

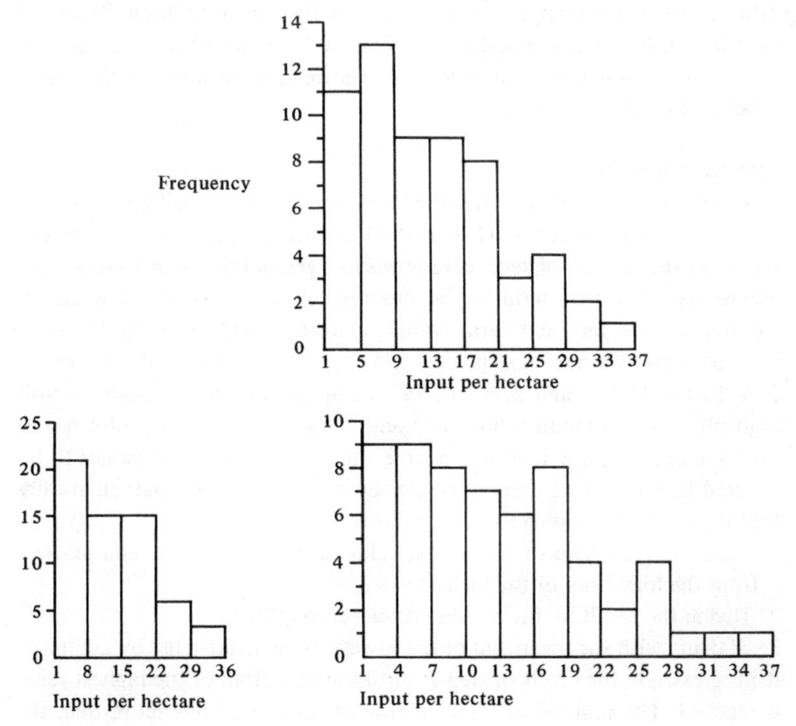

trary, although the simple formula mentioned above (p.56) can be a helpful guide. A histogram containing nine classes is used to illustrate the distribution in Fig. 5.4. Class limits are defined as 2–5, 6–9, 10–13, 14–17, and so on. Note the relationship between these limits and the class interval of 4, and also the class boundaries scaled along the horizontal axis. The histogram reveals that the distribution in question has a clearly defined positive skew. This characteristic is also shown by the other histograms in Fig. 5.4, which contain five and twelve classes respectively. Note, however, that the degree of generalization obviously increases as the number of classes is reduced, and the use of only five classes obscures some important detail.

Although the largest histogram contains no empty classes, it provides little more detail than the one with nine classes. This appears to support the result obtained from the formula that nine classes might be considered a practical maximum for a data set of this size.

The construction of a histogram requires the use of a fixed interval, but this method is not always the best suited to other forms of data presentation, such as the mapping of distributions. Its advantage lies in its ease of calculation, but it is not sufficiently flexible to cope with strongly skewed distributions, when the majority of values are concentrated within a fairly small part of the overall range.

Variable intervals

Class limits based on an interval which varies in some regular way are the most difficult to calculate. One method of introducing an interval of this sort involves the use of successive terms in a **geometric progression** as class boundaries. The first term in the progression is multiplied by a chosen amount to give the next term, which in turn is multiplied by the same amount to give the next in sequence, and so on. For example, the sequences 2, 4, 8, 16, 32, . . . and 2, 6, 18, 54, . . . are geometric progressions with multipliers, or **common ratios**, of 2 and 3 respectively. The problem is to find a common ratio that will enable the requisite range of values to be covered in the desired number of classes. It can be shown mathematically that the method of doing this is as follows:

1. Find the logarithm of the lowest value in the distribution, and subtract it from the logarithm of the highest value.
2. Divide the result by the number of classes required.
3. Starting with the logarithm of the lowest value, make a list by adding to it progressively the result of step 2, until the logarithm of the highest value is reached. The number of items in this list should be one more than the number of classes used.
4. Find the anti-logarithms of each of the terms in the list and use these as the class boundaries required. They should all be terms in a geometric progression, although the common ratio is unlikely to be a whole number.

If we consider the distribution listed in Fig. 5.1, the lowest value is 2, and the highest is 36. We may now work through the procedure outlined above.

1.

$$\begin{aligned} \text{Log. } 36 &= 1{\cdot}5563 \\ \text{Log. } 2 &= \underline{0{\cdot}3010} \\ \text{Log. } 36 - \text{Log. } 2 &= 1{\cdot}2553 \end{aligned}$$

2. Assuming that, for this example, there are to be only five classes,

$$\frac{1{\cdot}2553}{5} = 0{\cdot}2511$$

3, 4.

	Logs.	Anti-logs. (to 1 decimal place)
	0·3010	2·0
+ 0·2511	0·5521	3·6
+ 0·2511	0·8031	6·4
+ 0·2511	1·0542	11·3
+ 0·2511	1·3052	20·2
+ 0·2511	1·5563	36·0

The class boundaries are thus successive terms of the geometric progression 2, 3·6, 6·4, 11·3, 20·2, 36, . . . , in which the common ratio is about 1·8. Since we are using discrete data, we may define the class limits as 2–3, 4–6, 7–11, 12–20, and 21–36. With continuous data they would obviously become 2–3·5$\dot{9}$, 3·6–6·3$\dot{9}$, 6·4–11·2$\dot{9}$, and so on.

This method of defining class limits is well suited to distributions which exhibit a positive skew, since classes are narrow at the lower end of the range, where the majority of the values lie, but become broader towards the upper end where there are relatively few values. This results in a more balanced allocation of values to classes, a characteristic which is often helpful in mapping. Fig. 5.5, although not strictly a histogram, illustrates the way in which the pattern of class frequencies has been altered by adopting a variable interval.

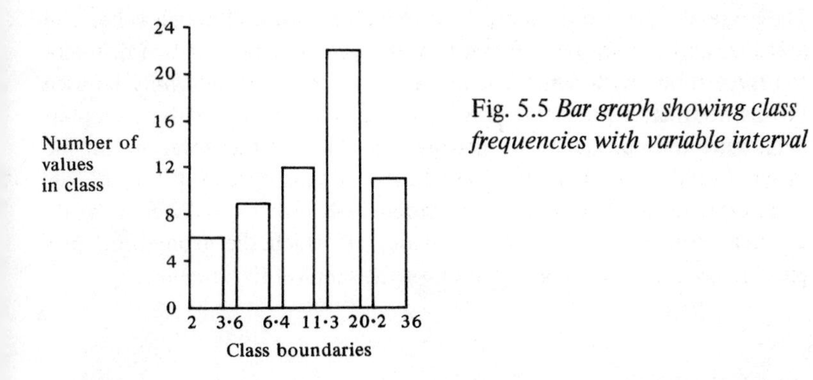

Fig. 5.5 *Bar graph showing class frequencies with variable interval*

Changes of this nature in the distribution of class frequencies may be referred to as **data transformations**. These are sometimes used to convert skewed distributions into approximately normal ones, for purposes of statistical analysis. In this case, a distribution with positive skew has been converted into one with negative skew, indicating that the strength of the transformation has been greater than required. A better result might be obtained by applying the same technique, but using square roots instead of logarithms to establish the class boundaries. This would produce a different transformation of the original data, thereby altering the frequency of values in each class. Square roots have a rather weaker effect than logarithms, and we might therefore expect that the new distribution would be fairly symmetrical about a central modal class, thus resembling a normal distribution.

Although the success of the transformation is not critical to the classification of data, it is convenient to be able to use alternative transformations in cases where a geometric progression of class boundaries is not suitable, for instance when the lowest value in the distribution is 0 (there is no logarithm of 0). If the distribution has a negative skew to start with, it is best to use squares, since these will produce class limits that become narrower towards the upper end of the range.

The calculation of class limits for any of the alternative transformations mentioned is no more difficult than the example given above. It involves working through the same four steps as before, and substituting terms as appropriate from the list below:

Transformation to be used	*For logarithm read*	*For anti-logarithm read*
squares	square	square root
square roots	square root	square
reciprocals	reciprocal	reciprocal.

These are the most commonly used transformations, although others are possible, and it may be necessary to use a combination of two different transformations with some sets of data. Reciprocals should only be used for distributions with a large positive skew, and then only with continuous data, since they produce very narrow class limits at the lower end of the range. If they are used, 'highest' should also be substituted for 'lowest', and vice versa. Each of these transformations will produce a different mathematical progression of class boundaries, of which the geometrical progression obtained by using logarithms is the simplest to recognize.

Which method?

Of the methods described in this chapter, the scattergram and cumulative frequency graph are probably the most useful for classifying geographical data. They offer a high degree of flexibility in the choice of class limits, since the class interval can vary irregularly, and not as a term in formal progression. Limits chosen in this way are more likely to achieve the essentials of a successful classification, as stated above (p. 55).

This is not meant to imply that the other methods covered are not used by geographers, for there are many examples of them in atlases, textbooks, etc. **Whichever method is chosen, it must be remembered that classification is not an end in itself. The final decision over both the number of classes, and their limits, should reflect the purpose for which the classification is made.**

Exercises

Fig. 5.6 *Size and functional content for service centres in East Anglia. Populations are taken from the 1961 census reports. Functional indices are based upon the number and type of functions in each centre*

Name of centre	Functional index	Population	Name of centre	Functional index	Population
Bury St. Edmunds	1233·31	21 179	Walsham-le-Willows	20·19	791
Newmarket	672·94	9636	Woolpit	20·16	963
Stowmarket	520·46	7795	Exning	16·93	1591
East Dereham	438·45	7199	Castle Acre	15·97	872
Diss	423·14	3681	Rickinghall Inferior	15·31	312
Thetford	376·70	5399			
Downham Market	364·60	2835	Botesdale	13·87	487
Wymondham	301·34	5904	Elmswell	13·57	1177
Mildenhall	253·02	7132	New Buckenham	13·15	375
Swaffham	250·34	3202	Hilgay	13·11	1343
Watton	217·88	2462	Mendlesham	11·78	933
Attleborough	205·69	3027	Stoke Ferry	11·67	723
Brandon	131·27	3344	Southery	10·71	1209
Eye	96.60	1583	Hopton	10·19	371
Harling	56·90	997	Stanton	9·60	1252
Hingham	51·93	1388	Old Buckenham	8·72	854
Feltwell	42·95	3192	Hockwold-cum-Wilton	8·47	848
Lakenheath	41·72	4512			
Fordham	36·85	1709	Mattishall	7·82	940
Ixworth	34·99	940	Cheveley	7·71	1624
Methwold	34·21	1560	Badwell Ash	7·13	331
Shipdham	29·36	1237	Bacton	7·06	528
Isleham	26·39	1392	North Lopham	6·82	461
Haughley	22·26	978	Barrow	6·80	856
Rickinghall Superior	21·01	369	Stowupland	6·64	1070
			Marham	6·58	3021
Kenninghall	20·77	782	Hockering	5·80	345

1. Attempt a classification of the service centres listed in Fig. 5.6, using the method which appears most suitable to you. Do not use more than four

G. W. Aschmann

classes. Compare the results you achieve using population with those based on functional index. What differences are there? How many centres are in each class? What names might you give to the classes you have established?
2. Attempt a transformation of the fertilizer input data given in Fig. 5.1, using square roots instead of logarithms. Use five classes, and obtain your class limits by working through the procedure described. Construct a bar graph showing the result of your transformation, and comment on the differences between this and the bar graph in Fig. 5.5.

Chapter 6

Mapping distributions

To the geographer, a numerical distribution is most meaningful when placed within a spatial context. Each item within the distribution then has two properties, one being its value (or quantity), and the other its location, which may be either a point or an area. Together, these properties produce the pattern of areal variation which constitutes a spatial distribution.

The description of a spatial distribution is best achieved, not by attempting to summarize the data statistically, but by illustrating the properties mentioned in the form of a map. Map construction is therefore a basic geographical technique, especially so since visual description is a vital first step in seeking to explain a distribution of this type. Two broad categories of maps may be distinguished:

a) **Qualitative maps.** These illustrate non-numerical distributions, which may be either continuous (geology, land use) or discontinuous (settlement, communications, lakes, rivers, coalfields, national parks). They show location, extent, and type–but not numerical value. Their construction is fairly straightforward once the necessary information has been collected, the chief requirement being a suitable scheme of colouring or shading to distinguish between items of different types.
b) **Quantitative maps.** These show variation in value or quantity over an area, and require greater preparation. Several types are available, the three most commonly used being the **dot map**, the **choropleth map**, and the **isopleth map**.

This chapter is concerned only with the methods involved in the construction of the quantitative maps mentioned above. Each of these is adaptable to a fairly wide range of data, but inevitably each has its own specialities, and its own strengths and weaknesses. To provide a basis for comparison, each will be used to illustrate the distribution of population in part of Norwich, given the data shown in Fig. 6.1. (N.B. These data were collected for a pre-metrication survey; the techniques illustrated are applicable to all data, whether in metric or Imperial units).

Fig. 6.1 *The distribution of population in part of Norwich. For each zone both total population and population density (persons per acre) are indicated, the latter being shown in italics*
(**Source:** Norwich Area Transportation Survey, 1969)

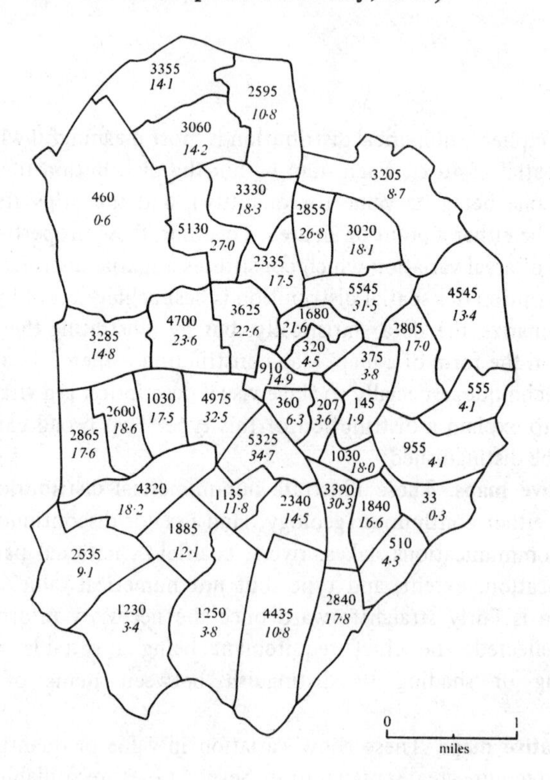

Dot maps

Some distributions, such as those of population, livestock, or a particular type of crop, relate not to measured values, but to numbers of items—people, cattle, acres—counted over a given area. If the total number of items is small, and the exact location of each is known, then it may be possible to represent the distribution as a point pattern (see Chapter 3). In many cases, it is neither practical nor necessary to do this, since the detail is either not available, or not required on the finished map. The dot map allows the major features of the distribution to be illustrated without showing the exact location of every individual item. Its principle is to use a dot of fixed size to represent the approximate location of a given number of items.

Although the construction of a dot map involves no major difficulties, its effectiveness can be influenced by a number of factors, which require careful consideration beforehand.

i) Enumeration areas

The area under study must be divided into a number of smaller sub-areas within which items are counted. This is roughly equivalent to the grouping of data within numerical distributions, and has the same implications. Although the number and size of the areas used is frequently decided independently by the nature of the units for which information is available (counties, parishes, farms, etc.), it is important to remember that a few large areas provide less information than many smaller ones.

ii) Dot value

This is simply the number of items which each dot represents on the map, and dot value is therefore closely related to the total number of dots used, as expressed by the relationship

$$\text{Dot value} \times \text{total no. of dots} = \text{total no. of items.}$$

Normally the dot value is chosen as a convenient round number, and the number of dots worked out to the nearest whole number from this, but in some cases it may be preferred to decide how many dots to use, and thus to derive the dot value. For example the use of 100 dots would ensure that each dot represented 1% of the total number of items in the study area. If possible, the value chosen should be low enough to ensure that even enumeration areas with few items receive one or two dots, thus avoiding an impression of total emptiness on the finished map. However, if the range of area totals is large, too low a value may result in the need for excessive numbers of dots to represent the larger quantities. In such cases some sort of compromise usually has to be made to simplify the construction of the map.

iii) Dot size

This must be chosen with regard to dot value and the scale of the map. The best visual impression is normally achieved when the size is such that dots just begin to merge in the areas with the highest density of items. However, if a relatively small number of dots is being used on a large scale map it may be impractical to achieve this effect without making the dots awkwardly large.

iv) Dot location

In the absence of more detailed information, dots must be distributed evenly within each enumeration area. Some dots should be placed on or close to the boundaries between areas, so that these border zones do not stand out as artificial gaps on the completed map. In some cases dots may be subjectively placed on the basis of further information about the distribution of items within individual areas, such as the location of a village within a parish, or of certain soil conditions on a farm (which may influence livestock or crop distributions).

The steps involved in the construction of a dot map are best explained by illustration using the data given in Fig. 6.1. Suppose that we wish to use a dot map to represent the distribution of population in this part of Norwich.

1. **Delimit enumeration areas**, and count the total number of items within each. This has been done in Fig. 6.1. In this case the information has been taken from a previous survey.

2. **Count the total number of items** to be represented on the map. The total population of the 44 enumeration areas is 104 925, or approximately 105 000.

3. **Decide the dot value.** The figures illustrate the problem which arises when the range of area totals is large, in this case from 33 to 5545. A dot value of about 30 is obviously impractical, for this would require the use of 3500 dots, and so it must be accepted that a few areas must remain empty of dots if a realistic value is to be chosen. For purposes of illustration a dot value of 500 will be used.

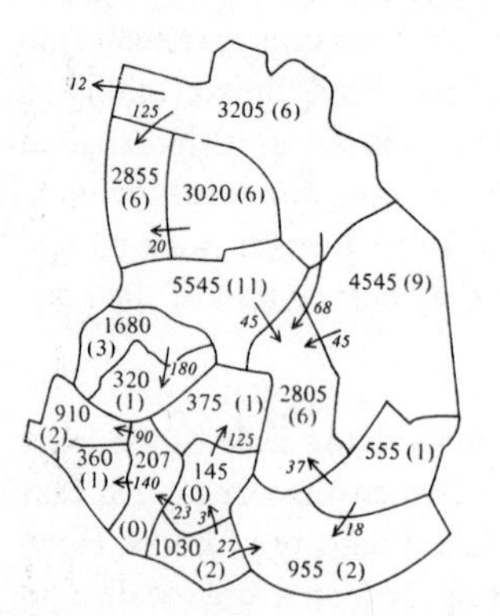

Fig. 6.2 *Adjustments to enumeration area population (part of map only). Adjustments are shown by a figure and arrow. The numbers in brackets indicate the dot requirement in each area*

4. **Allocate dots to areas.** Once the dot value has been decided, it is possible to work out the number of dots required in each area. A problem which will always arise here is that the number of items in an area is rarely a perfect multiple of the dot value, and a certain amount of redistribution of items between areas is needed to achieve this. This is done by trial and error, and the adjustments can be recorded (preferably in pencil to allow for mistakes) on the base map, as illustrated in Fig. 6.2.

The process of redistribution is inevitably rather arbitrary, and some care is needed to keep alterations to a minimum, by working steadily across the map. It is as well to check afterwards that the total number of dots allocated is correct; with a total population of 105 000 and a dot value of 500, there should be 210 dots.

5. **Decide the dot size, and locate dots within each area.** This is the final stage in the construction of the dot map, and may require some experimentation to produce the best result. For this reason it is a good idea to use tracing paper placed over the base map, and to work initially in pencil. The completed map of the distribution of population in Norwich is shown in Fig. 6.3.

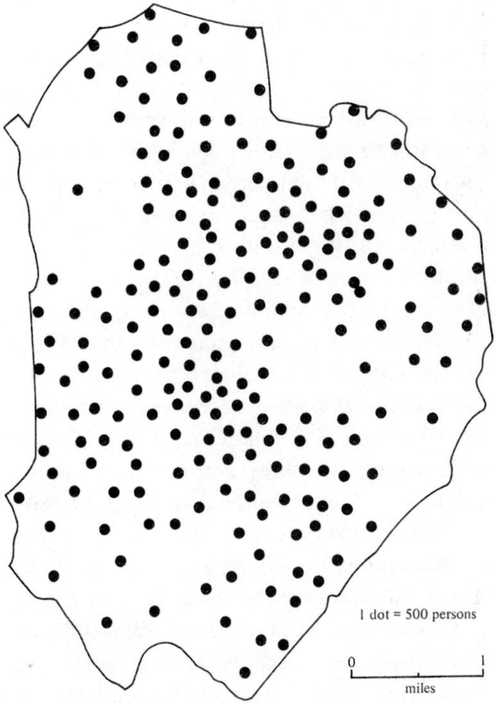

Fig. 6.3 *The population distribution represented by a dot map*

The effectiveness of the finished map is limited by two factors. Firstly, although the dots are large enough to stand out clearly, they do not begin to coalesce in the most densely populated areas. To achieve this dot size could be increased, but a better solution might be to use a smaller dot value (say 250), and thus increase the number of dots, which for purposes of illustration has been kept rather small. Secondly, and more important, an unfortunate impression of uniformity is created by the even spacing of dots. The major features of the distribution are blurred, not by the mapping technique but by the lack of detailed information; for this reason dots have to be evenly spaced. In these circumstances the dot map can only indicate general variations in population density between areas, with the disadvantage that it is not suited to quantitative interpretation. **If nothing is known of the distribution of items within enumeration areas, it can be strongly argued that other techniques are both easier, and able to provide better results, than the dot map.**

Choropleth maps

The principle of the choropleth map is a very simple one, in which straightforward techniques of shading or colouring are used to build up a pattern of spatial variation between areas. To eliminate the effect of area size on the quantities involved, each total is expressed in standardized form as a ratio to area. Thus population totals become densities of persons per square mile (or some other unit of area), and crop acreages are converted into percentages of the total area of farmland. Other types of information normally measured within an areal context can also be represented on a choropleth map. Examples are land value, crop yield, or fertilizer input, all of which are commonly measured in units per acre or hectare.

Areas are allocated to classes on the basis of the values relating to them, the distribution of areas belonging to each class being shown on the map by the system of shading or colouring adopted. Thus the basic problem in constructing a choropleth map is one of classification, and in this respect the technique is simply a practical extension of the methods discussed in Chapter 5. The procedure can be illustrated by using the Norwich population data, this time referring to the area densities shown in Fig. 6.1.

1. Obtain a map showing area boundaries, and **mark on the standardized values calculated for each area.** This stage is represented by Fig. 6.1.

2. **Divide the range of values into classes.** The values should be considered as an ordinary numerical distribution, since their spatial distribution is of no significance at this stage. The number of classes used, and the choice of

class intervals, should reflect the considerations discussed in Chapter 5, although a further practical limitation on the number of classes may be imposed by the need for each class to be visually distinctive on the finished map. In this respect colouring is generally more flexible than black and white shading, and thus allows the use of a greater number of classes, though both techniques can lose their effectiveness if too many variations are required.

If the total number of classes has to be kept fairly small, it is obviously important that class limits should be chosen carefully, so as to reflect the character of the data, and to achieve the maximum possible variation between classes. This will ensure that the finished map effectively illustrates the most important areal contrasts. An irregular class interval is generally most convenient in this respect, since it allows greater flexibility in the choice of limits. One way of investigating the character of the data is to construct a **cumulative area graph**, such as that illustrated in Fig. 6.4.

As its appearance shows, this graph is a version of the cumulative frequency graph, but differs in that cumulative area, rather than frequency, is

Fig. 6.4 *Cumulative area graph for Norwich population data*

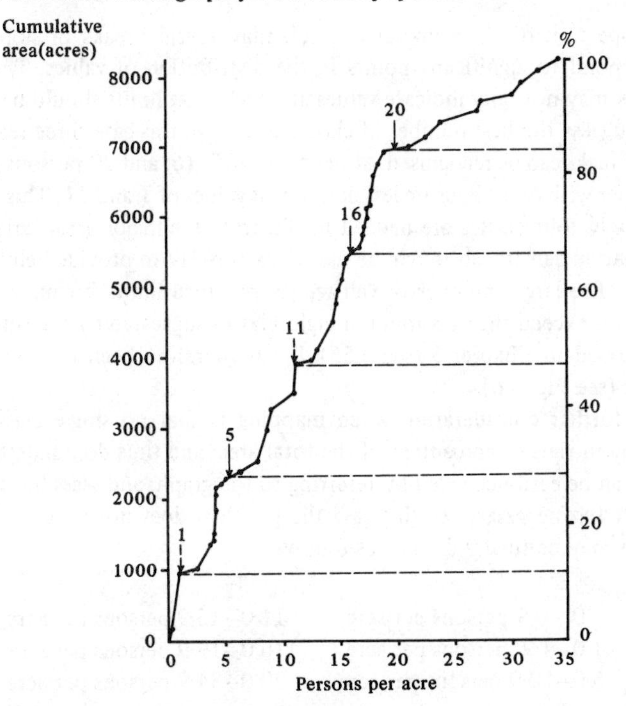

plotted along the vertical axis. The method of construction is very similar, requiring the tabulation of data as shown in Fig. 6.5.

Fig. 6.5 *Method of data tabulation for cumulative area graph*

Population densities (ranked from lowest to highest)	Acreage of area to which density refers	Cumulated acreage
0·3	123·2	123·2
0·6	779·2	902·4
1·9	72·9	975·3
3·4	358·7	1334·0
3·6	71·3	1405·5
3·8	98·7	1504·4
3·8	333·0	1837·0
etc.		

Density is then plotted against the appropriate cumulated area, expressed either as an actual quantity, or as a percentage of the total area. The construction of suitable scales allows both to be shown on the same graph (see Fig. 6.4).

Inspection of the completed graph may reveal breaks of slope which correspond to significant points in the distribution of values. Thus these breaks may not only indicate values at which class limits should be set, but also suggest the best number of classes to use. In this case three reasonably clear breaks can be recognised, at densities of 5, 16, and 20 persons per acre together with two which are less definite, at values of 1 and 11. This suggests that only four classes are needed to illustrate the major areal variations in population density, although the use of six is likely to provide helpful extra detail. The larger number is well within practical limits for mapping, and does not exceed the maximum of eight classes suggested by the simple rule mentioned in Chapter 5 (see p.55). It has therefore been used in this example (see Fig. 6.6).

A further consideration when mapping is that no single class should occupy too large a percentage of the total area, and thus dominate the map. This can be easily checked by referring to the graph, and class limits can be adjusted as necessary. In this case the problem does not arise, and the six classes may be finally defined as follows:

0– 0·9 persons per acre	11·0–15·9 persons per acre
1·0–4·9 persons per acre	16·0–19·9 persons per acre
5·0–10·9 persons per acre	20·0–34·9 persons per acre

It is not at all essential to construct a cumulative area graph in order to define class limits. In some cases a simpler alternative, such as the scattergram (see p.57), may be preferred. With many sets of data the sizes of individual enumeration areas may not be readily available, and the construction of a cumulative area graph is therefore impracticable. Whichever method is used, it is as well to check that a reasonable balance is maintained between the areal extent of different classes on the finished map.

3. **Choose a scheme of shading or colouring to represent classes on the map.** The main object here is to achieve a grading from lighter tones, to represent values at the lower end of the range, to darker tones for the higher values. Although colouring has greater flexibility, and is usually preferable to black and white shading, there are times when the latter must be used, for instance if the finished map is to appear in one-colour print. Effective results can be produced by careful variation of the ratio of black to white in the shading

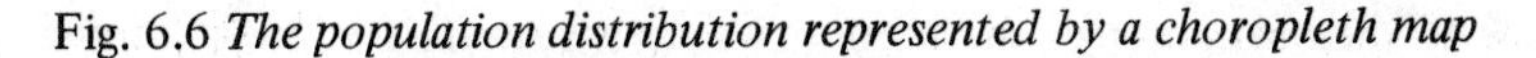

Fig. 6.6 *The population distribution represented by a choropleth map*

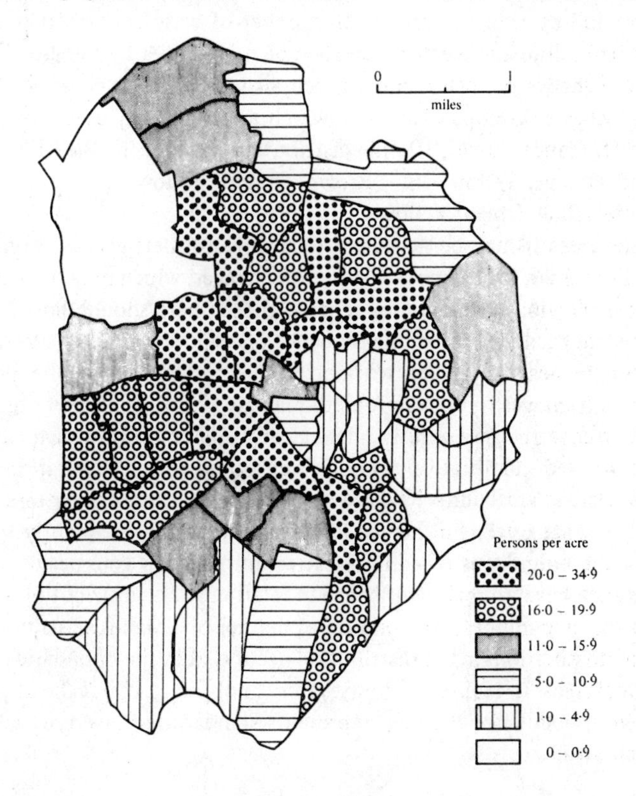

pattern, by changing the width and spacing of lines, or the density of dot stippling. Solid black shading is frequently used at the top end of the range, while at the lower end areas are often left unshaded, but it is not essential to employ either of these two extremes. In this respect the degree of contrast required in the finished map must be suggested by the data themselves. If necessary it is possible to arrange the shading so that there is a greater contrast between some classes than others, reflecting the character of the distribution. The system of shading adopted for the Norwich data is shown with the completed map in Fig. 6.6.

If colours are being used, selection should reflect the order of colours within the spectrum, viz. red, orange, yellow, green, blue, indigo, and violet. In general softer colours such as yellow and green (from the middle of the spectrum) should be used for the lower values, while areas of high value can be emphasized by using the more striking colours at either end of the spectrum, such as red, blue, and violet. Gradual variation of tone from one class to the next through the range of values is best achieved by colouring lightly, and by using a fairly small number of basic colours. Blending between one colour and the next can then be made easier by the use of several different shades of each colour, which also provides extra variation in the colour scheme. Groups of colours which are commonly used (for instance in the Ordnance Survey 10 mile distribution maps of the British Isles) are:

a) Red, Orange, Yellow or Brown, Orange, Yellow
b) Violet, Blue, Green, Yellow.

In some cases it may be necessary to distinguish between two contrasting groups of values, for example population changes, which may be positive or negative. Obviously the colour scheme chosen here should emphasize the contrast, normally by using one sequence of colours for positive values, and another for negative values, working from different ends of the spectrum.

The advantages of the choropleth map are that it is relatively quick and simple to construct, presents the data clearly and in a form which can easily be interpreted quantitatively, and is well suited to the illustration of data which relate to areal units. Its disadvantages arise because the pattern of area boundaries has a strong influence on the finished map. Frequently these are simply the boundaries of administrative units such as counties or parishes, and as such have no real significance in relation to the distribution itself. In effect the map implies two conditions, neither of which is actually true:

a) that no variations in the distribution occur within the boundaries of individual areas. In reality, of course, this is not the case, but the impression is inevitable since the map represents standardized 'average' values for each area;

b) that abrupt changes in the distribution occur along the boundaries between areas belonging to different classes. In fact, changes are most likely to be gradual, and to bear little or no relation to the boundary lines. Exceptions may occur however; for instance, the level of fertilizer input per hectare may be expected to vary from one farm to another, although other factors are also involved (e.g. soil quality).

Despite these limitations, the choropleth is probably the most commonly used type of quantitative map. Because of its disadvantages, it is important that attention should be given to overall impression, rather than to detailed characteristics. As Fig. 6.6 illustrates, it can then provide an effective and reasonably accurate picture of the distribution to which it relates.

Isopleth maps

Maps of this type are drawn on the same principle as the contour map, which belongs to the same general category. **Isolines**, or 'contours', are constructed so as to connect points of equal value, and the nature of the distribution is illustrated by the pattern of these lines on the finished map. The distribution of values which can actually be measured at a given point (such as rainfall, temperature, and of course altitude) is commonly represented in this way, but the technique can also be used with data which relate to areas, rather than points. This is achieved by locating a point within each area, giving it the appropriate value, and using it to represent the area as a whole. It is therefore possible to illustrate the population density data in this way, and the steps involved in the construction of an isopleth map for this distribution are set out below.

1. **Construct a map showing the distribution of values as 'spot-heights'**. If, as in this case, the values relate to areas, it is usual to locate the points centrally within each area. If it is known that the distribution within a particular area is uneven, then it may be desirable to locate the point off-centre towards that part of the area which has the higher value. Fig. 6.7(a) illustrates this first stage in construction for part of the Norwich area.
2. **Decide upon the number of isolines to be used, and their values.** This is the equivalent of choosing a contour interval for a relief map, except that in this case the interval need not be constant. Isolines are in effect class boundaries, and should therefore be chosen in accordance with the classification techniques dealt with in Chapter 5. In this respect the procedure is exactly the same as that used in the classification of data for the construction of a choropleth map. With physical distributions (relief, temperature, rainfall, etc.), it is customary to use a fixed isoline interval, but with human

distributions (population, land value, etc.) it is often better to let the data themselves suggest which values to use. Values have already been selected for the Norwich population data, and isolines will therefore be drawn at values of 1, 5, 11, 16, and 20 persons per acre.

3. **Draw in the isolines on the map.** This is done by interpolation from the point values plotted in step 1. It is best to work initially in pencil, since the process is essentially one of trial and error, and alterations will almost certainly have to be made.

Fig. 6.7 *Stages in the construction of an isopleth map. Part of the Norwich area is illustrated. The stages are: A location of central 'spot-heights' for each area; B plotting of isoline positions between adjacent spot-heights; C, D drawing of isolines; these two maps represent alternative solutions, both of which fit the available information*

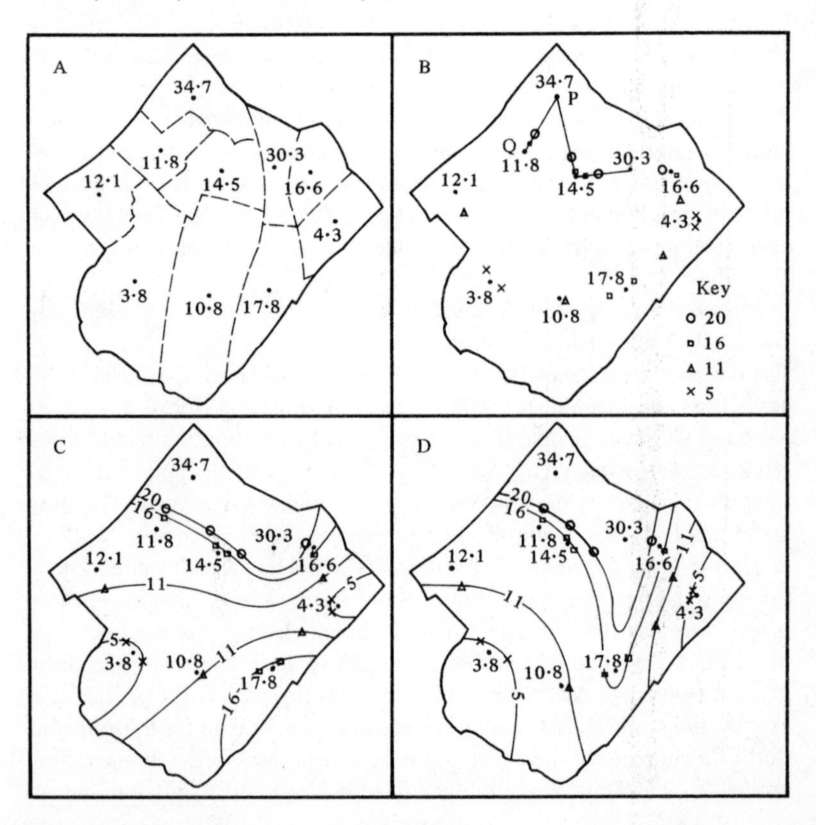

Very few, if any, of the existing point values will correspond to the chosen isoline values, so it is first necessary to add extra control points to the map, through which it is known that the isolines must pass. The location of these points is deduced from the pattern of values already on the map. In Fig. 6.7(b), for example, the value of point *P* is 34·7, while that of the neighbouring point *Q* is 11·8. Between these two values lie the chosen isoline values of 16 and 20. Assuming a constant gradient between *P* and *Q*, it is a fairly simple matter to estimate the location of the necessary control points. Thus the value 20 must lie just over one third of the distance between the two points from *Q*. This distance can be measured, but it is far quicker, and only slightly less accurate, to fix the intermediate positions by eye. The diagram illustrates the location of other control points derived in this way for values of 5, 11, 16, and 20.

It is now possible to sketch in isolines between the appropriate control points. This is not always a simple operation, especially if the density of known values over the map is rather low. A common problem is illustrated in Figs. 6.7(c) and (d), which represent two alternative interpretations of the distribution based on the same control points. Uncertainty arises here because the centre of the map lacks a known point value, and surrounding points are too far apart to provide conclusive evidence. In such a situation the decision made must be rather subjective, but it often occurs that one solution appears to fit better into the general pattern of the distribution. In this case the layout shown in Fig. 6.7(d) has been chosen for this reason and is included in the complete isopleth map (see Fig. 6.8).

Occasionally, two or more alternatives may appear equally feasible, in which case it is possible to estimate a 'missing' value by a simple rule-of-thumb method. This involves taking the known values of surrounding points, calculating their mean, and then locating this value centrally within the area which they enclose. This method is rather arbitrary, and cannot guarantee an accurate answer, but if applied consistently it is probably better than taking pot luck.

4. **Ink in and label the isolines**, once their final pattern has been decided. All other information pencilled in during the construction of the map can now be rubbed out. The completion of this stage is represented by the finished map in Fig. 6.8.

5. This stage is optional, and involves the shading or colouring of areas between the isolines in a manner akin to the contour colouring of relief maps, in order to provide greater visual impact. Since the isolines represent class boundaries, this operation is the equivalent of area shading on the choropleth map, and the same principles are applicable (see above).

Fig. 6.8 *The population distribution represented by an isopleth map. The isoline values are in persons per acre*

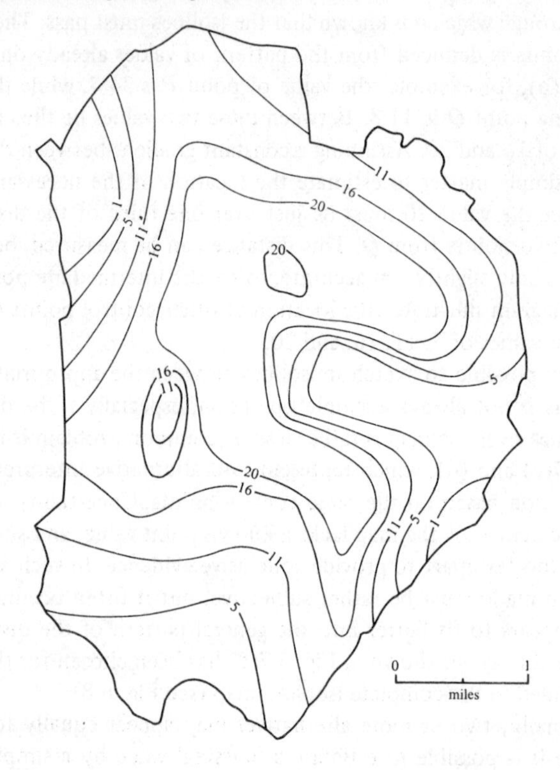

Which method?

Clearly the criticisms made of the choropleth do not apply to the isopleth map. Its main advantages are that data relating to areas can be represented without the inclusion of area boundaries (true also of the dot map), and that changes in value are shown to occur smoothly and continuously, rather than in abrupt steps. As a result, the isopleth is useful for illustrating the general trends within a distribution. Like the choropleth, it can be interpreted quantitatively, and it is also the most flexible of the three maps used in terms of the types of data which it can represent.

On the other hand, the isopleth is certainly more difficult and time-consuming to construct than the choropleth map, although it is probably easier than the dot map. There is also a greater subjective element involved which introduces a degree of uncertainty over the detailed accuracy of the finished map, especially if there is a shortage of known values. This is most likely to be a problem with distributions which by nature are patchy and uneven, and which do contain abrupt changes from one area to another, thus making general trends more difficult to identify. These conditions are not uncommon with human distributions, and in such cases the choropleth is probably the better map to use.

Fig. 6.9 *Population statistics for the counties of England and Wales.* (**Source**: 1971 Population Census, Preliminary Report)

County	A	B	County	A	B
W. Suffolk	24·5	1·0	Cornwall	10·2	1·1
Huntingdonshire	24·3	1·6	Surrey	9·8	5·9
Berkshire	23·1	3·4	Lincs (Lindsey)	9·6	1·2
Oxfordshire	21·0	2·0	Staffordshire	9·5	6·2
Essex	20·6	3·7	Norfolk	9·4	1·2
Bedfordshire	19·3	3·8	Lincs (Kesteven)	8·9	1·2
Buckinghamshire	19·3	3·0	Cambridgeshire	8·7	1·4
W. Sussex	17·8	3·0	Yorks (N.Riding)	8·7	1·3
Northamptonshire	16·3	2·0	Devon	8·6	1·3
Hertfordshire	15·9	5·6	Westmoreland	8·0	0·4
Flintshire	15·7	2·6	Nottinghamshire	7·7	4·5
Rutland	15·7	0·7	Gloucestershire	6·5	3·3
Hampshire	15·7	4·0	Denbighshire	6·0	1·1
Kent	15·4	3·7	Herefordshire	5·5	0·6
Anglesey	14·5	0·8	Derbyshire	4·5	3·4
Dorset	14·2	1·4	Monmouthshire	3·7	3·3
Wiltshire	14·0	1·4	Pembrokeshire	3·3	0·6
Isle of Wight	13·3	2·9	Warwickshire	3·1	8·2
Somerset	13·1	1·6	Yorks (E.Riding)	3·0	1·8
Shropshire	12·4	1·0	Yorks (W.Riding)	2·7	5·2
Leicestershire	12·3	3·6	Lincs (Holland)	2·2	1·0
Cheshire	12·0	5·9	Cardiganshire	2·2	0·3
E. Sussex	12·0	3·5	Glamorgan	2·1	5·9
Worcestershire	10·6	3·8	Caernarvonshire	0·9	0·8
E. Suffolk	10·4	1·7	Durham	0·4	5·5
Lancashire	– 0·5	10·5	Carmarthenshire	– 3·4	0·7
Cumberland	– 0·8	0·7	Breconshire	– 3·6	0·3
Radnorshire	– 1·1	0·1	Greater London	– 7·9	46·7
Montgomeryshire	– 3·2	0·2	Merionethshire	– 8·2	0·2
Northumberland	– 3·2	1·5			

A – % change in population, increase or decrease, 1961–71.

B – population density, persons per hectare, 1971.

Fig. 6.10 *Ordnance Survey map extract of Norwich. The solid line marks the boundary of the area covered in Fig. 6.1*

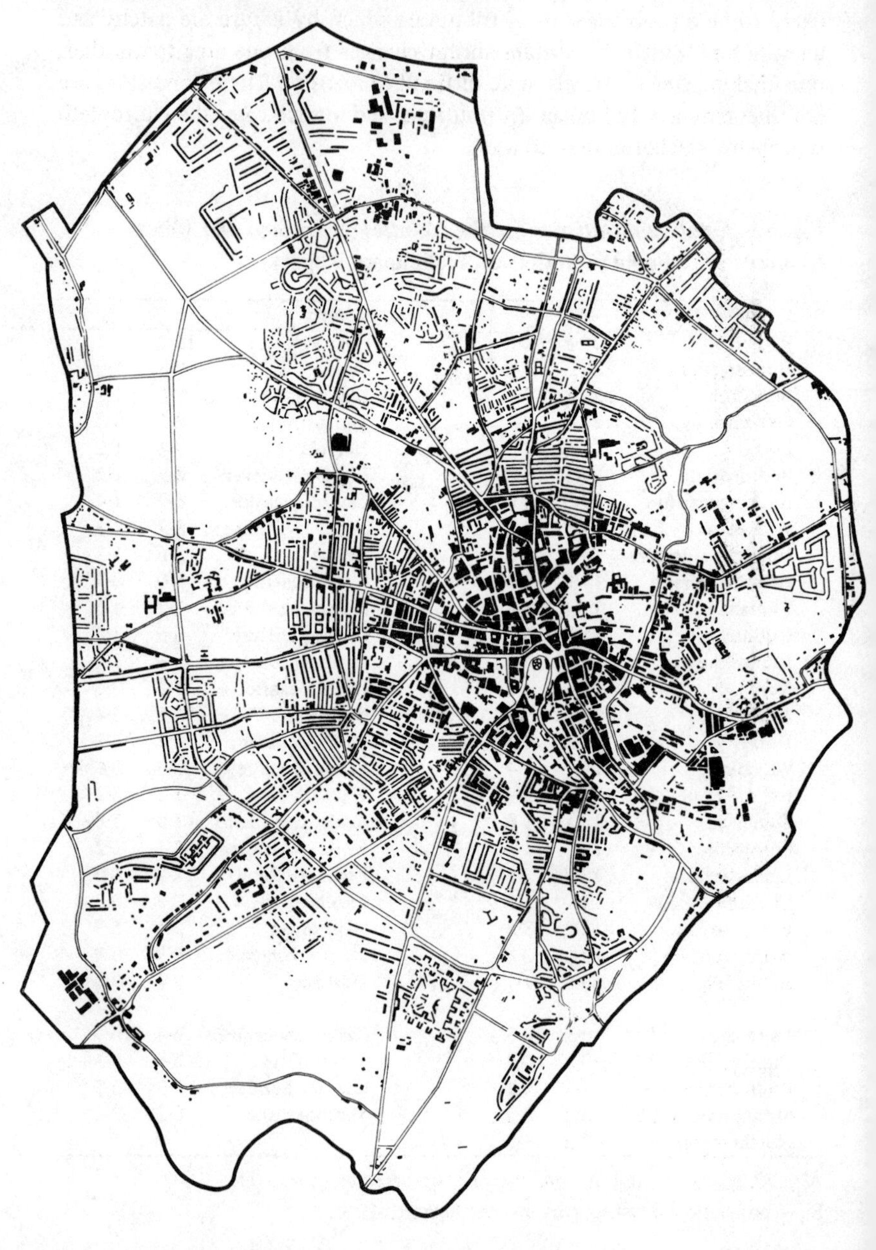

The three finished maps produced by the methods explained in this chapter may differ superficially in appearance, but each one clearly illustrates the same basic distribution of population. In practice, for this and any other distribution, the geographer must decide which type of map is likely to be most effective in combining the properties of accuracy, clarity, simplicity, and visual impact. He must therefore consider both the nature of the data which he has to represent, and the various pros and cons of each of the methods available to him. In this respect the mapping of spatial distributions is no different from the summarization of numerical distributions.

Exercises

1. Construct suitable maps to illustrate the two sets of data given in Fig. 6.9. In each case justify your choice of map, and comment on the effectiveness of the result.
2. Trace a number of parish boundaries from a 1″ Ordnance Survey map of your local area. If you live in an urban area you may be able to use wards. Find out the population of each of the parishes or wards you have selected by referring to the census report for your county. Construct a dot map to portray as accurately as possible the distribution of population in the area chosen. Use your local knowledge and information on the O.S. map to help you to locate the dots.
3. Using the O.S. map extract of Norwich given in Fig. 6.10 to help you, construct a dot map to illustrate the distribution of population. Base your map on the data given in Fig. 6.1, but use a smaller dot value than in the example given (say 200–250), and locate the dots with regard to the infor-

Fig. 6.11: *Non-manual workers as a percentage of total employed population for enumeration areas in Norwich (see Fig. 6.12)*
(**Source:** Norwich Area Transportation Survey, 1969)

Area	%	Area	%	Area	%	Area	%	Area	%
A1	28·2	B3	40·4	D3	76·5	E4	88·1	G1	11·1
A2	49·7	B4	24·8	D4	22·0	E5	77·3	G2	29·0
A3	41·9	C1	33·3	D5	48·5	F1	25·4	G3	48·6
A4	14·2	C2	32·0	D6	36·4	F2	29·3	G4	51·1
A5	26·4	C3	87·0	D7	20·5	F3	22·3	G5	56·3
A6	19·3	C4	32·0	D8	20·4	F4	30·7	G6	74·3
A7	29·1	C5	51·6	E1	50·1	F5	44·2	G7	37·1
B1	57·1	D1	35·0	E2	76·9	F6	84·1	G8	32·9
B2	29·7	D2	39·1	E3	37·6	F7	49·1		

Fig. 6.12 *Norwich enumeration areas. See exercise 4 in this chapter, and exercise 4 in Chapter 7*

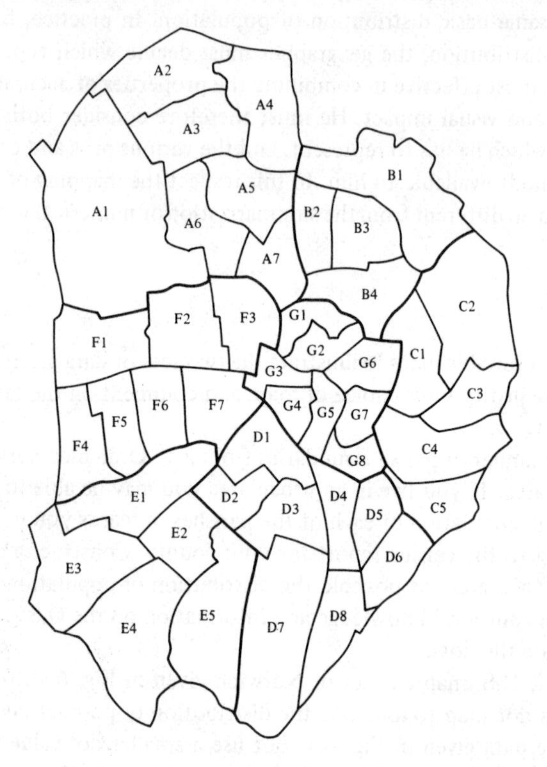

mation on the map extract. Compare the effectiveness of your result with that of the map in Fig. 6.3.

4. Fig. 6.11 lists the proportion of non-manual workers amongst the total employed population of each of the enumeration areas in Norwich indicated in Fig. 6.12. Use either a choropleth or an isopleth map to represent this data, giving reasons for your choice.

Further reading for Chapter 6

Cole, J.P., and King, C.A.M., *Quantitative Geography* (Wiley, 1968), pp. 192–200.

Dickinson, G.C., *Statistical Mapping and the Presentation of Statistics* (Arnold, 1963), Chapter 2, especially pp. 44–59.

Haggett, P., *Locational Analysis in Human Geography* (Arnold, 1965), pp. 214–18.

Monkhouse, F.J., and Wilkinson, H.R., *Maps and Diagrams* (Methuen, 1963), especially pp. 31–46, 222–35, 277–97, and 360–8.

Toyne, P., and Newby, P.T., *Techniques in Human Geography* (Macmillan, 1971), Chapter 3, especially pp. 85–92 and 96–101.

Chapter 7

Mapping flows

Geographical data quite frequently relate to items which are either moving, or being moved. Information of this sort may occur in one of several forms:

a) quantities passing through a series of checkpoints where counts are made or rates of flow measured (river discharge, pedestrian and vehicular traffic flow);
b) known quantities passing along a given section of route in a specified time (bus and train services);
c) quantities passing between known points of origin and destination over a given period (freight tonnage, population movements). These differ from (b) in that the actual route taken may not be specified.

The most important feature of distributions such as these is that variation in quantity occurs, and must therefore be represented, on a linear basis. Maps drawn to illustrate them must be able to show both the pattern of movement (usually governed by a network of routes or channels), and the quantity of flow along each route. There are only two methods commonly used for this purpose, these being the flow line and the desire line.

Flow line maps

A flow line is a line of variable width drawn to represent the quantity of flow passing along a given route; this route is indicated by the course which the line follows. Flow line maps illustrate the distribution of movement within an area by using flow lines of appropriate width for each section of the network on which movement is taking place. Their construction is fairly simple, and the procedure may be illustrated by using data on road traffic flow in N.W. London.

1. **Construct a map of the route network, and mark onto it the location of each checkpoint, or of each section for which the quantity of traffic is**

known. This stage is represented by Fig. 7.1, which shows part of the primary road network of N.W. London, together with the location of a number of countpoints for which traffic flow data are available (see Fig. 7.2). Note that count points are positioned on all routes close to main intersections, where the greatest changes in flow may be expected to occur. Most route sections thus have two count points, one at each end.

2. **Mark onto the map the total quantity of traffic recorded at each count point**. Alternatively, number the count points and tabulate the data as in Figs. 7.1 and 7.2. All figures marked onto the map should be in pencil, so that they can be erased later. Count points are unnecessary for scheduled flows such as bus and train services, but the number of services passing along each route section should be recorded in the same way (see Fig. 7.9).

3. **Decide upon a suitable scale of line width**. This will depend upon the range of quantities to be represented, the density of the route network, and

Fig. 7.1: *Part of the primary road network of N.W. London, showing the location of traffic count points. Note that the M1 motorway connects with the remainder of the road network at its southern end only*

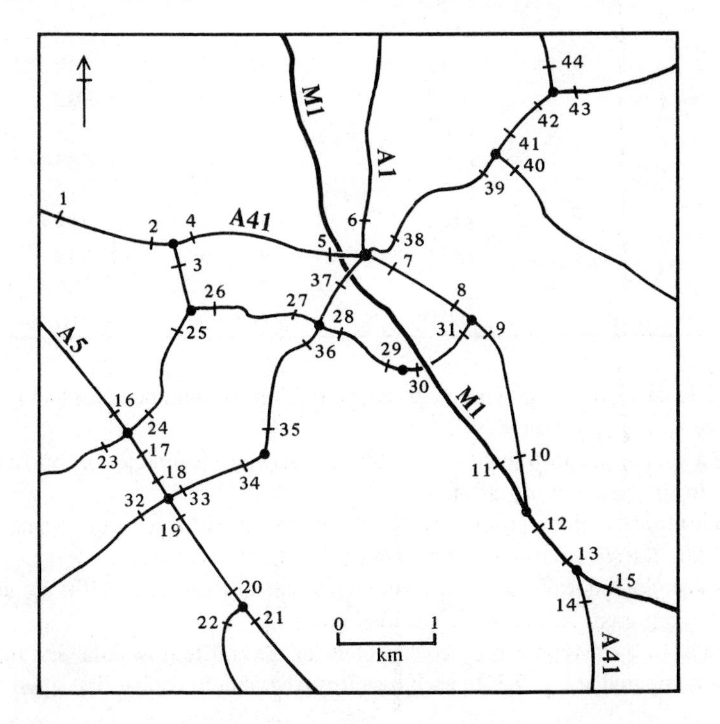

Fig. 7.2 *Daytime off-peak traffic flows past the count points shown in Fig. 7.1. The totals given are average hourly flows during the period 10.00 a.m. to 4.00 p.m.*
(**Source:** Surveys conducted by the Greater London Council, Department of Planning and Transportation)

Count Point No.	Vehicles per hour	Count Point No.	Vehicles per hour
1	1348	23	1053
2	1315	24	1116
3	380	25	995
4	1396	26	723
5	1403	27	580
6	1812	28	411
7	2299	29	587
8	2377	30	771
9	2518	31	825
10	2028	32	713
11	2010	33	655
12	4030	34	777
13	4450	35	866
14	2839	36	839
15	1642	37	998
16	1107	38	549
17	1563	39	563
18	1509	40	448
19	1224	41	713
20	1383	42	584
21	1372	43	585
22	605	44	210

the scale of the map. Three alternative methods should be considered, the choice lying between:

a) a simple proportional scale, in which line width is directly proportional to the quantity of traffic;
b) a more complex proportional scale, in which width is proportional to, say, the square root or logarithm of the quantity involved;
c) a graduated scale, in which a limited number of fixed line widths are used to represent flows within specified limits.

Scales of each type can be constructed for the traffic flow data, and these are compared in Fig. 7.3. In each case it is advisable to decide first upon the

Fig. 7.3 *Alternative scales of flow line width for the road traffic data. The scales are: (a) directly proportional; (b) proportional to the square root; (c) graduated. Note that only scale (a) gives a true visual impression of the relative sizes of different flows.*

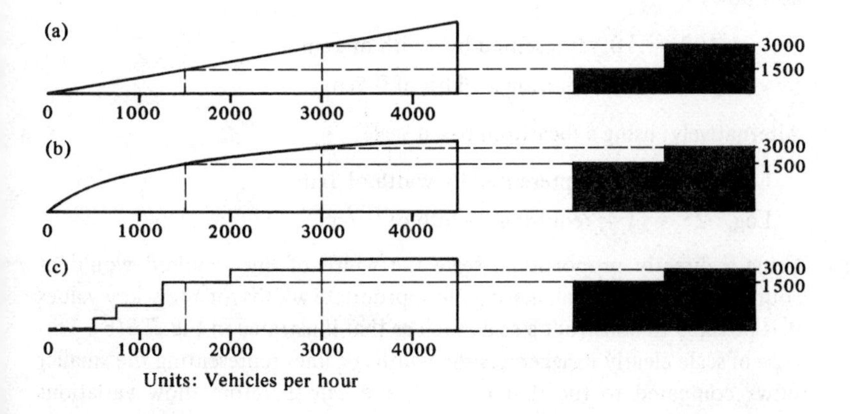

maximum width of line that can be used without reducing the effectiveness of the completed map. Obviously there should be no overlap between separate flow lines (except in the immediate vicinity of route nodes), but it is also important that there should be adequate spacing between each flow line and its neighbour to avoid the impression of overcrowding. Clearly the maximum width must be fixed with regard to the densest part of the route network, particularly if the largest flows also occur there.

The first type of scale (directly proportional) can be constructed very simply by deciding the maximum line width to be used, allocating this to the highest flow recorded (or the nearest round number above it), and completing a scale triangle. It permits quantitative interpretation of the sizes of individual flows, and also enables flows to be compared without reference to the scale triangle--since one flow line roughly double the width of another represents approximately twice the amount of traffic (see Fig. 7.3(a)). As a result the completed map gives a good visual impression of the relative sizes of different flows. On the other hand, difficulties may arise when the range of quantities to be represented is large, especially if there are many relatively small flows and a few much larger ones. In this situation, most of the flow lines must inevitably be narrower than is really desirable so that the few larger flows can also be represented. The finished map is therefore dominated by the broader flow lines, and it is possible for quite important variations amongst the smaller flows to be largely obscured.

The second type of scale provides a possible solution to the problems mentioned above. Suppose that the largest flow recorded is 100 and that this is to be represented by a flow line of 1cm. width. The width of line required to represent a flow of 25, using a square root-based scale, is given as follows:

$\sqrt{100} = 10$, represented by width of 1cm.
$\sqrt{25} = 5$, requiring a width of 0·5cm.

Alternatively, using a logarithm-based scale:

Log. 100 = 2·0, represented by width of 1cm.
Log. 25 = 1·4, requiring a width of 0·7cm.

Using a directly proportional scale the width of line required would of course be 0·25cm. By calculating the appropriate widths for a few key values it is possible to construct a scale such as that illustrated in Fig. 7.3(b). This type of scale clearly exaggerates the widths of lines representing the smaller flows compared to the first method, and can therefore show variations amongst them in greater detail. However, the advantages of direct proportionality are lost and, although this has no effect upon the accuracy of representation, it does create a misleading visual impression of relative size which is apparent in Fig. 7.3(b).

The graduated scale requires greater preparation, since the data must first be split up into classes. Those used for the scale shown in Fig. 7.3(c) were chosen by using a scattergram to illustrate the main size groups in the traffic flow data (see Chapter 5). Although the class interval may not be constant, line widths should increase steadily by equal amounts from one class to the next. Since each width must be clearly distinguishable from the one above or below it, the number of classes used must be kept fairly small. If necessary, very small flows can be represented by using a dotted or pecked line, as shown in the diagram.

A scale of this sort is useful for representing traffic data which exhibit a large range and uneven distribution of flow sizes. The use of a limited number of flow line widths simplifies the construction and interpretation of the finished map. However, the scale does not permit precise measurement of individual flows from the map, since each line width represents a range of flow sizes, rather than a single quantity. Another drawback, which results from the use of a variable class interval, is that flow line width does not increase in direct proportion to the range of flows within each class. In this respect diagram (c) is more similar to (b) than it is to (a). Thus a graduated scale provides a poor basis for comparison of the actual quantities of traffic passing along different routes. Despite this, the finished map gives a good visual impression of a route 'hierarchy'; with reference to the scale it is possible to discover the general quantitative relationship between one level of the hierarchy and the next.

Of the three types of scale mentioned, the first is probably the best for general use. However, the choice must depend upon the nature of the flow data and the purpose for which the flow map is drawn. The data given in Fig. 7.2 could quite easily be represented by the first method, but do exhibit both a fairly large range and a rather uneven distribution. Thus, for purposes of illustration, a graduated scale will be used in this case.

4. **Mark line widths onto the map.** This should be done for each count point or route section, in accordance with the scale chosen. Use a ruler, and make two small marks in pencil–one on each side of the route line. Position the marks so that this line is central between them, and their distance apart is as given by the scale.

5. **Draw in the flow lines.** Using the pencil marks as guides and following the general direction of the route, draw in the edges of each line, one section at a time. If the quantity of flow is constant throughout the section, line width will not alter from one end to the other, and the edges can be drawn parallel to the central line. However, data collected at count points will almost certainly exhibit discrepancies between one point and the next. These represent the net result of traffic gains and losses en route, not all of which can possibly be recorded. In such cases, any necessary changes in line width should be made at an intervening junction, if one exists. Failing this, line width may be altered in one of two ways, as illustrated in Fig. 7.4.

Fig. 7.4 *Flow line construction*

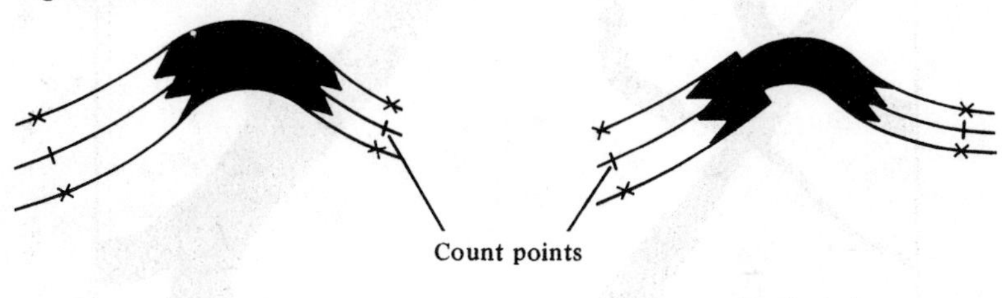

If a proportional scale is being used, width may be steadily increased or reduced from one point to the next, so that the flow line is slightly tapered. With a graduated scale an abrupt change in width must be made halfway between the count points, although of course this is only necessary if the quantities involved are in different classes. An obvious disadvantage in this case is the unrealistic impression given of an abrupt change in traffic flow at the halfway point.

Once drawn, flow lines are usually coloured-in to improve their visual impact. Black is commonly used for this purpose, since it contrasts well with the white background. All other marks made during construction of the map can now be erased. Where routes converge upon important centres, the flow lines representing radial traffic flows tend to merge and overlap. In such cases the appearance of the finished map can often be improved by drawing a circle of suitable radius around the centre, and not continuing the flow lines inside it.

The completed flow line map representing the traffic data for N.W. London is shown in Fig. 7.5. The graduated scale is the same as that used

Fig. 7.5 *The pattern of traffic flows in N.W. London represented by a flow line map*

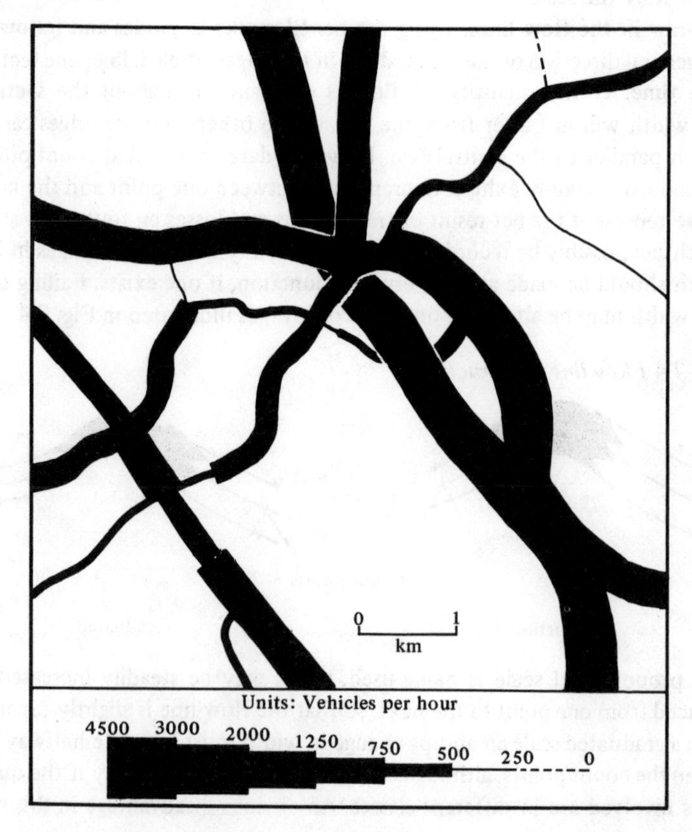

in Fig. 7.3(c), except that line widths have been reduced to a more practicable level. The map reflects the pros and cons of a graduated scale, for although a route hierarchy is clearly visible, a misleading impression of relative flow size is given. Careful reference to the scale is therefore necessary to provide a valid basis for comparison.

Two-directional flows

Flow data may sometimes distinguish between traffic moving in different directions along the same route. The flow line map can be adapted fairly easily to represent such data. This is illustrated in Fig. 7.6, which shows the pattern of pedestrian movement in central Norwich. The diagram on the left presents the flow data from which the flow line map on the right has been constructed. In this case a simple proportional scale has been used, since the range of flow sizes is fairly small. Two flow lines are of course needed for each route section, instead of one, and arrows are suitably incor-

Fig. 7.6 *The pattern of pedestrian flows in central Norwich: an example of a two-directional flow line map. The quantity of flow in each direction is indicated on the left-hand diagram street by street. The data are based on the aggregate of three separate counts, each of five minutes' duration*

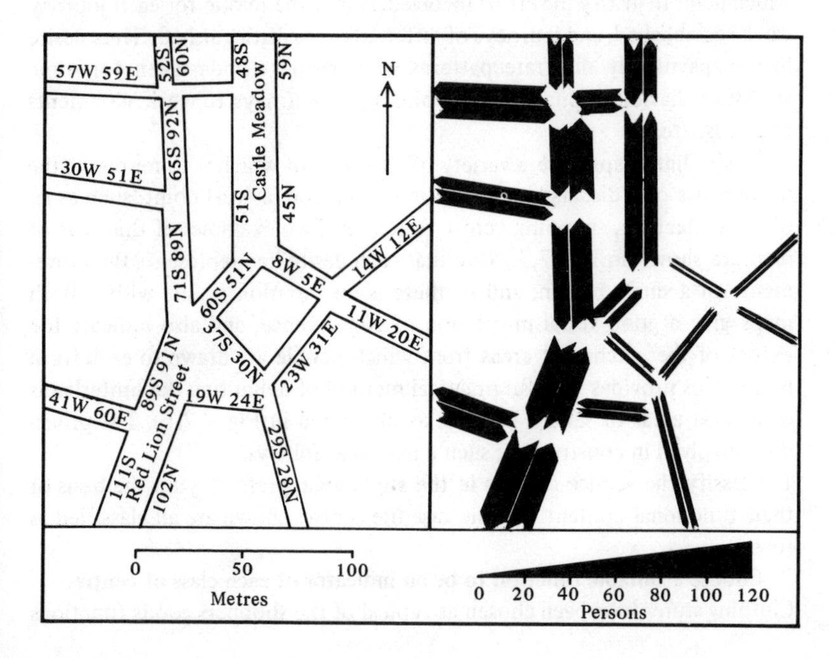

porated to indicate the direction of movement which each represents. The flow lines are broken up into separate sections to eliminate the overlapping which would otherwise occur at junctions, thus simplifying the appearance of the finished map. A method for taking pedestrian counts is given in S.I.G. 2, Appendix B.

Desire line maps

Desire lines are similar to flow lines in that their width represents a given quantity of movement. The main difference between them is that a desire line is drawn straight between known points of origin and destination, and thus takes no account either of the actual route followed, or of the type of transport used. It represents diagrammatically the need for, and general direction of, movement between two points, rather than the flow of traffic along part of a route network.

Clearly the information required for the construction of a desire line map is more difficult to collect than that needed for flow lines. Details of the origin and destination of each individual journey are of a qualitative nature, and cannot be observed and counted in the same way as traffic flows on a road network. In fact, the information often has to be collected at first hand by time-consuming questionnaire methods. One obvious advantage of this which helps to justify the effort involved, is that the reason for each journey can be established, and journeys of different types distinguished. Thus desire line maps usually illustrate patterns of movement undertaken for some specific behavioural purpose (shopping trips, journeys to work, shipments of goods, etc.).

Desire line maps have a variety of uses, one of which is to represent the movements of a dispersed population to and from a focal point, such as an office or factory, shopping centre or school. Two examples of this type of map are shown in Fig. 7.7. Note that each desire line represents the movements of a single person, and so there is no variation in line width. Both maps give a good visual impression of convergence, and also indicate the extent of the catchment areas from which people are drawn to each focal point. This provides a useful practical method of delimiting the hinterlands or market areas of service centres, as illustrated in Fig. 7.7(b). The procedure involved in constructing such a map is as follows:

1. **Classify the service centres in the study area**, preferably on the basis of their functional content. In this case the centres shown are all classified as towns.

2. **Choose a suitable function to be an indicator of each class of centre.** Clothing stores have been chosen as typical of the shoppers goods functions

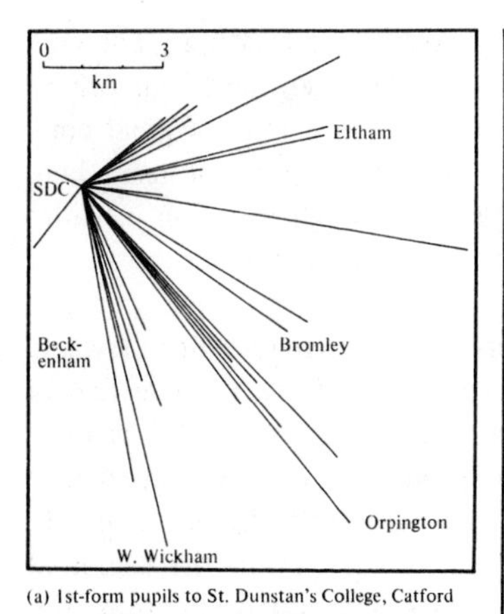

(a) 1st-form pupils to St. Dunstan's College, Catford

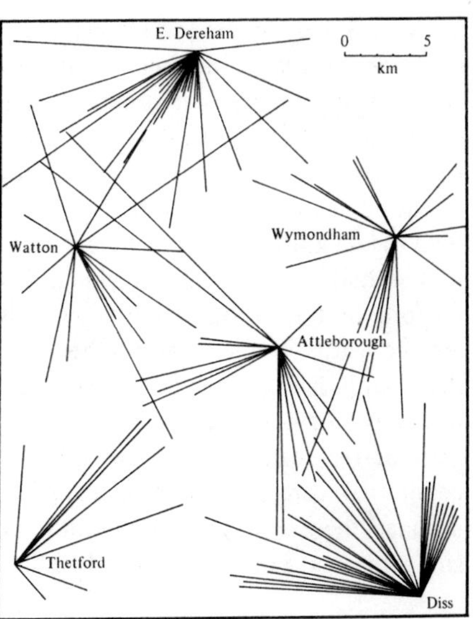

(b) Shoppers to clothing stores in East Anglia

Fig. 7.7 *Desire line maps used to delimit catchment areas. For further examples of maps of this sort see Berry, B.J.L.,* Geography of Market Centres and Retail Distribution *(Prentice-Hall, 1967), Chapter 1*

in towns; butchers might be chosen as indicators for the convenience functions of villages.

3. **Conduct a sample questionnaire survey to discover the pattern** of shopping trips to service centres of a given class. Stand outside a shop of the chosen type, and question a fixed number of customers (say 50) in each centre to find out where they have come from. Alternatively, question people living in the areas between the centres and find out where they last went to purchase the appropriate items. The first method is simpler, more reliable (the shoppers are actually seen in the centre), and usually quicker but may not give sufficient areal coverage. In fact, many of the shoppers questioned may live in the centre itself. The second method, used in this case, is less likely to leave gaps in coverage, but takes longer and depends upon accurate answers being given.

4. **Plot each trip as a desire line to build up the finished map**. The pattern of desire lines should indicate the approximate position of the 'breaking point' between one centre and the next, and thus allow the hinterland of each centre to be defined. (An explanation of the calculation of 'breaking point' is given in S.I.G. 1, Chapter 1, Fig. 1.8).

The use of desire lines to represent individual journeys as shown in Fig. 7.7 is only practicable if the total number of journeys is limited by the use of a small sample, and there are few possible destinations. The construction of a large number of individual lines between numerous origins and equally numerous destinations would result in a map of great complexity and little practical value. Furthermore, the property of desire lines to represent different quantities of movement through variation in width is wasted. Thus individual desire lines are seldom used for purposes other than the delimitation of catchment areas.

The half dozen or so individual desire lines which join Bromley and Catford in Fig. 7.7(a) could obviously be replaced by a single line of greater width. If the total number of journeys to be represented is large, those with similar origins and destinations must be grouped together in this way for the reasons given above. The grouping should take into account the scale of the map, which in many cases limits the degree of detail that can be used in actually plotting origin and destination. For instance, there would be little point in attempting to plot the exact starting and finishing points of individual journeys between two towns 50km

Fig. 7.8 *A desire line map showing inter-regional migration in England and Wales. Each line represents the quantity of net movement in the direction indicated during a 12-month period in 1960–1*
(**Source:** 1961 Population Census–figures are based on a 10% sample survey)

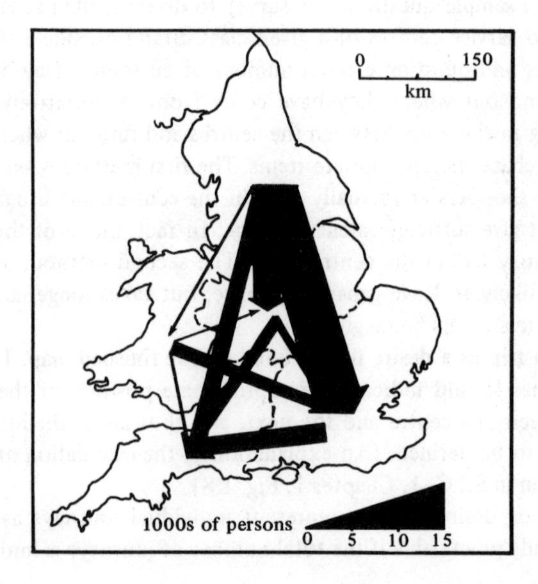

apart, whereas the total quantity of movement between them can be represented quite simply by a single desire line.

In Fig. 7.8 a desire line map is used to represent the pattern of migration in England and Wales. In this case five large regions provide a basis for grouping journeys, and the number of desire lines is thus well within practicable limits. The use of smaller areas, such as counties, would clearly result in far too complicated a pattern of lines, and the effectiveness of the map would be greatly reduced. The approximate centre of each region is used as a terminal point for desire lines, and arrows indicate the direction in which the net balance of migration occurs. The 'drift to the south' is particularly apparent.

Despite the similarity between flow line and desire line maps, they are not direct alternatives. This is obvious if we refer back to the different types of data distinguished at the beginning of the chapter. Data of type (a) require the use of a flow line map, and cannot be represented in any other way, since nothing is known of the origin and destination of individual journeys. Movements of category (c), on the other hand, can only be represented by a desire line map unless the route taken is specified. Even then a true flow line map cannot be constructed, since all other movements along the route are ignored. It is only information of type (b) that can be properly represented by either map, since both the terminal points and the total quantity of movement along each route are known. In this case the map may illustrate either **the overall pattern of traffic (flow line)**, or **the availability of services between given termini (desire line)**. Thus the type of map used depends ultimately, as with the mapping of static distributions, upon both the nature of the data and the purpose for which it is drawn.

Exercises

1. Construct a flow line map using a proportional scale to illustrate the data given in Fig. 7.2. Compare your finished map with the one in Fig. 7.5, which uses a graduated scale. Which scale gives the better result in your opinion?
2. Using a suitable scale of your own choice, construct a flow line map to illustrate the passenger train service data given in Fig. 7.9.
3. Construct a desire line map to represent the data given in the following table:

Number of week-day main-line passenger train services between the cities named (in both directions)

	Birmingham	Cardiff	Manchester	Sheffield
London	74	51	36	31
Birmingham	–	10	14	29
Cardiff	–	–	5	4
Manchester	–	–	–	33

What problem would arise if more cities were included on the map?

4. The table below gives the volume of traffic movement in Norwich between the large enumeration areas marked in Fig. 6.12. The figures, which are in passenger car units (p.c.u.s), are based on average week-day

Fig. 7.9 *Numbers of inter-city train services operating on a normal week-day over part of the British Rail main line network*
(Source: British Rail passenger timetables, 1972)

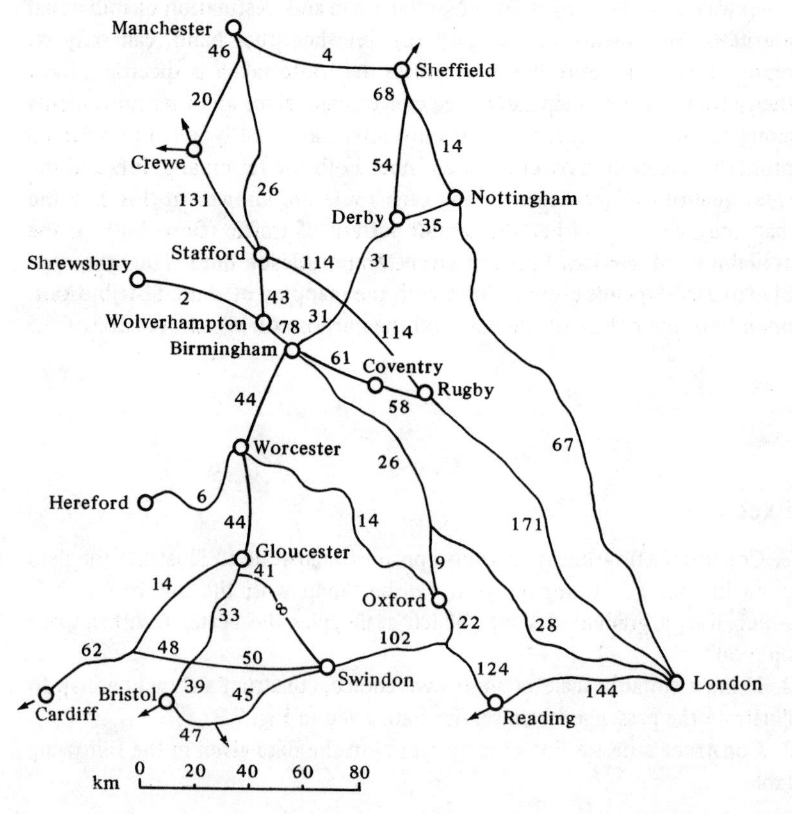

movement between the hours of 0700 and 1900. The p.c.u. values used were: car–1·00; light commercial vehicle–1·00; other commercial vehicle–2·00; motor-cycle–0·75; private coach–3·00.

	and	B	C	D	E	F	G
between	A	2283	1669	2258	690	3049	6351
	B		1641	1218	499	1523	3897
	C			2308	627	1609	4633
	D				2288	2812	6368
	E					1490	3873
	F						6622

(**Source**: Norwich Area Transportation Survey, 1969)

Construct a desire line map to illustrate the pattern of movement revealed by these figures.

Further reading for Chapter 7

Dickinson, G.C., *Statistical Mapping and the Presentation of Statistics* (Arnold, 1963), pp. 63–5.

Everson, J.A., and FitzGerald, B.P., *Settlement Patterns* (Longman 1969), page 99, and *Inside the City* (Longman 1972), pp. 95, 143–4.

Monkhouse, F.J., and Wilkinson, H.R., *Maps and Diagrams* (Methuen 1963), especially pp. 254–7 and 370–3.

Toyne, P., and Newby, P.T., *Techniques in Human Geography* (Macmillan, 1971), pp. 94–6.

Chapter 8

Symbols and graphs

In addition to the maps discussed in Chapters 6 and 7, several other methods of presenting data in visual form are available to the geographer. Only those which are most commonly used will be dealt with here, since these are generally the simplest and most effective. There are basically two ways in which the data may be illustrated:

a) as **symbols** drawn to represent different quantities, which can be positioned on maps to show the locations of the points or areas to which they relate;
b) as **graphs** which represent data by scaling value or quantity along one or more axes.

The range of possible applications for both these methods is very large, and between them they are suitable for use with almost all forms of geographical data. The examples given can only illustrate the techniques of construction, rather than the full scope of application, of each of the methods discussed.

Quantitative symbols

These are generally, but not always, used to represent data which must be considered within a spatial context, since each symbol can be individually located on a map. Several types are available, but circular symbols are most commonly used.

Proportional circles

Fig. 8.1 illustrates both the location and population size of the major settlements in Oxfordshire. Superficially it resembles a dot map (see p.69), but in this case the symbols are variable in size, and can be used to represent different quantities, instead of a fixed number of items. Symbols of this type are called proportional circles, since they are drawn so as to be proportional in area A to the quantities Q that they represent. In math-

Fig. 8.1 *Size and location of settlements in Oxfordshire shown by proportional circles*

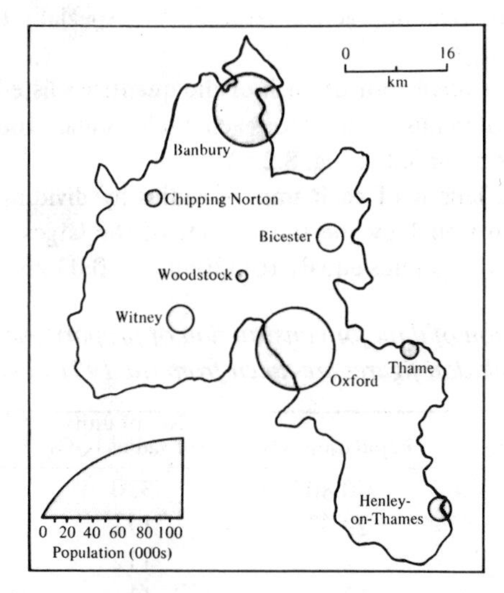

ematical terms we can express this as A ∝ Q (A varies with Q) and since the symbols are circular we also know that $A = \pi r^2$, where r is the radius of a circle of area A. It follows that $\pi r^2 \propto Q$. As π is a constant, and the relationship is one of proportionality rather than equality, we can ignore π, and simply write $r^2 \propto Q$. This means that $r \propto \sqrt{Q}$, and the radius of each circle should thus be proportional to the square root of the quantity it represents.

Considered in practical terms, the result obtained means that if we wish to represent two quantities, one of which is double the other, the radii of the circles drawn should differ by a factor of $\sqrt{2}$, or 1·41. Thus the radius of the larger circle would be less than half as much again as that of the smaller circle, but its area would be double.

Fig. 8.2 lists the populations of the settlements represented by proportional circles in Fig. 8.1. To calculate the radii of the circles needed for each of these quantities, we must work through the following procedure.

1. **Decide upon the maximum size of circle that can be used** to represent the largest quantity. This must obviously depend upon the scale of the map, and the density of settlements within the area. In this case it is clearly desirable to maintain the separate identity of each centre, and circles should neither overlap, nor contact each other. A radius of about 10mm might be considered the upper limit if Oxford is not to merge with Woodstock. In

some cases, where several large quantities occur close to each other, it may be impossible to avoid some degree of overlap without reducing the size of the smaller circles to an impractical level. This is especially true if the range of quantities is large.

2. **Calculate the square root of each of the quantities listed.** This gives the number of units of radius required for each circle. Square roots of the population figures are recorded in Fig. 8.2.

3. **Calculate the length of each unit of radius** by dividing the maximum radius decided in step 1 by the square root of the largest quantity. Thus, in this case, 1 unit of radius equals 10/330mm, or 0·03mm.

Fig. 8.2 *Tabulation of data for construction of proportional circles used in Fig. 8.1. Population figures are taken from the 1971 Census*

Name of settlement	Population (Q)	No. of units of radius ($\sqrt{Q}$)	Actual radius (r)
Oxford	108 805	330	9·9mm
Banbury	29 387	171	5·2mm
Witney	12 552	112	3·4mm
Bicester	12 355	111	3.3mm
Henley-on-Thames	11 431	107	3·2mm
Thame	5948	77	2·3mm
Chipping Norton	4761	69	2·1mm
Woodstock	1961	44	1·3mm

4. Calculate the actual radius to be used in each case, by multiplying the appropriate number of radius units by the length arrived at in step 3. Thus the circle used to represent Bicester should have a radius of 111 × 0·03 = 3·3mm. Radii have been calculated and are listed in Fig. 8.2.

If the number of circles required is large, the calculation of radii in this way becomes rather laborious. A useful short-cut in such a situation is the construction of a continuous scale of circle size. This scale, which may be called a **nomograph**, indicates the appropriate radius needed to represent any given quantity (within a specified range). Radii for a number of key values must be worked out to begin with, but all other radii required can then be read off the scale without further calculation.

The first step in the construction of a nomograph is to decide the maximum radius to be used, and to calculate the length of each unit of radius in the manner described above. Next select suitable values within the range of quantities to be represented, and find their square roots. The values chosen should be perfect squares, and must be fairly evenly spread through the range. In the case of the Oxfordshire data, we may take values of 2500, 10 000, 22 500, 40 000, 62 500 and 90 000, which have square roots of 50, 100, 150, 200, 250 and 300 respectively.

The finished form of the nomograph is shown in Fig. 8.3. This is obtained by plotting the selected quantities against their square roots, and drawing in a smooth curve through the points located in this way. Square roots are scaled as units of radius along the vertical axis, which should be marked off accurately according to the length of each unit previously calculated. Besides providing the correct radius for any quantity within the range of the horizontal scale, the graph also serves as the basis for a key to the sizes of proportional circles. It is only necessary to double the vertical scale to obtain the diameter corresponding to any given quantity. It is then a simple matter to compare the diameter of any circle on the map with the key, and read off the quantity it represents (see Fig. 8.1).

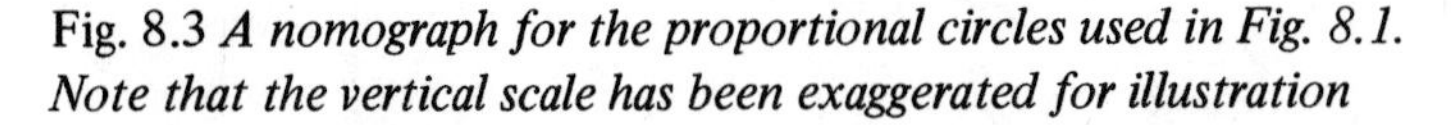

Fig. 8.3 *A nomograph for the proportional circles used in Fig. 8.1. Note that the vertical scale has been exaggerated for illustration*

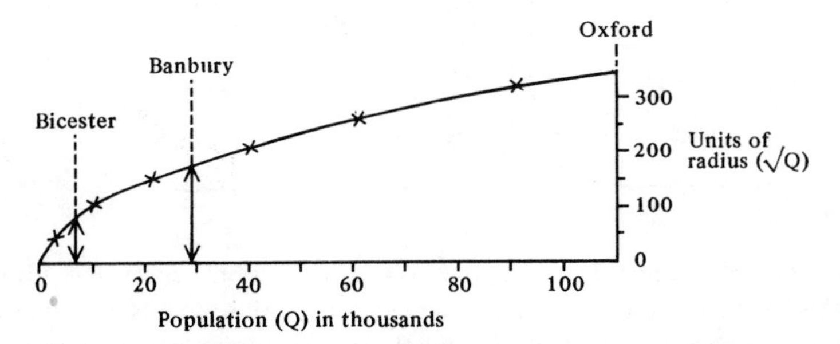

In Fig. 8.1 each proportional circle is placed so as to indicate the location of the centre to which it relates. Although the population of each centre is of course distributed over an area, the scale of the map is such that each circle may be regarded as having a location at a point. In some cases, however, each quantity may be measured over a larger area (for example, population within a county), and circles must then be located centrally within each area. This is illustrated in Fig. 8.4, where proportional circles are used to represent the output of coal from different mining areas in the British Isles. Note that area boundaries must be marked in when symbols of this type are used.

A range of other proportional symbols exists, including bars, squares, segments of circles, cubes, and spheres. Each of these is proportional either in length, area, or volume to the quantity which it represents, and each can

Fig. 8.4 *Proportional circles used to illustrate coal production in Britain by National Coal Board areas. The outputs shown are those for the year 1970–71* (**Source**: Figures published by the National Coal Board)

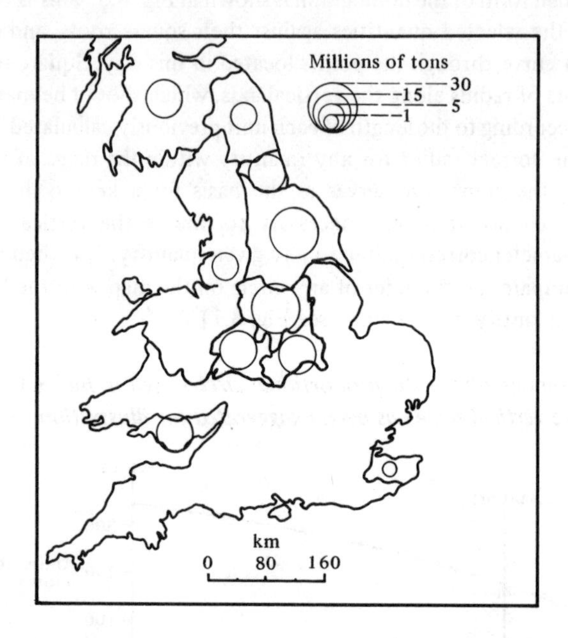

be located on a map in the same way as a proportional circle. The advantages of the circle lie in its fairly simple construction, and its neat appearance on finished maps. It is particularly suitable for use with data relating to points, since it focuses attention on the location of each point. Because of its shape it may be used in conjunction with dot symbols to represent areas of high density on dot maps, provided that dot size is in this case made proportional to dot value (see p.67). The only major disadvantage of the proportional circle is that it is possibly more difficult to estimate differences in area (and thus quantity) by comparing the sizes of circles than it is if some other types of symbol are used.[1]

Divided proportional circles (pie graphs)

In many cases the quantities represented by proportional circles can be subdivided into a number of component parts. Coal production statistics may be available for different types of coal, population totals may be split up into age or income groups, and acreages of agricultural land may be separated into different land use types. Such data can be illustrated by dividing each circle into a number of sectors, each occupying an appropriate share of the total area of the circle.

Fig. 8.5 *Working population characteristics of three towns. The totals are broken down into 5 socio-economic groups, defined as follows:*
A professional workers, employers, managers; B foremen, skilled manual workers; C non-manual workers (other than A); D semi-skilled and unskilled manual workers, agricultural workers; E others (including armed forces) (**Sources:** 1966 10% sample population census)

	Cambridge		Brighton		Coventry	
Group	Number	%	Number	%	Number	%
A	5880	19·9	8370	15·6	13 380	11·5
B	9600	32·5	20 580	38·4	52 170	44·7
C	6550	22·1	11 130	20·8	17 040	14·6
D	6750	22·8	12 100	22·6	32 280	27·6
E	800	2·7	1390	2·6	1880	1·6
Total	29 580	100·0	53 570	100·0	116 750	100·0

Fig. 8.5 lists the socio-economic characteristics of the populations of Cambridge, Brighton, and Coventry. In each case the total population given is that of both economically active and retired males over the age of 15. Each total is subdivided into five different socio-economic groups, and the percentage of the total in each group is also indicated. Clearly there are differences between the three towns, both in the size of the total population and its distribution amongst the various groups. Divided proportional circles, or pie graphs as they are often called, provide a useful means of representing this data visually, as illustrated in Fig. 8.6.

The first stage in construction is to draw proportional circles to represent the total population of each town. The next step is to subdivide each circle into sectors which, by their size, represent the percentage of the total contained within each sub-group. Clearly a 90° sector occupying a quarter of the circle would represent 25% of the total quantity, a 180° sector 50%, and so on. For any given percentage we can calculate the interior angle of the sector required simply as the percentage of 360°. Taking the sector for Cambridge, group A, as an example:

$$\text{interior angle required} = \frac{19{\cdot}9}{100} \times 360° = 71{\cdot}6° \ (72°).$$

Fig. 8.6 *Pie graphs used to represent the data given in Fig. 8.5*

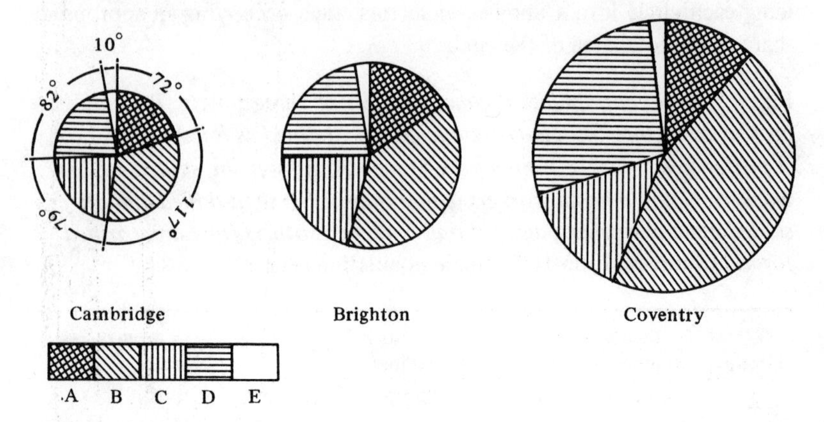

In general terms, the angle for a sector which represents $x\%$ of the total quantity is given by $x/100 \times 360°$, or simply $x \times 3{\cdot}6°$. Thus we have only to multiply each of the figures listed in the % columns of Fig. 8.5 by 3·6 to arrive at the angles required.

The north (0°) line is normally used as a starting point for division of the circles, and sectors are marked in progressively in a clock-wise direction. It is as well to check before starting that the angles calculated add up to 360°, and small adjustments may have to be made to allow for the effects of rounding off to the nearest whole number. In some cases it may be desirable to begin with the largest sectors and work downwards in size, but this is not essential. In fact the order is not critical, provided that it is the same for each circle. Once marked in, different sectors should be distinguished by a suitable scheme of shading or colouring (see Fig. 8.6).

Fig. 8.7 *Divided proportional bars used to represent the data given in Fig. 8.5. These are simpler to construct than pie graphs, and may be preferred if map representation is not required. The lengths of the bars, and of their subdivisions, are proportional to the quantities they represent. See Fig. 8.6 for key to shading*

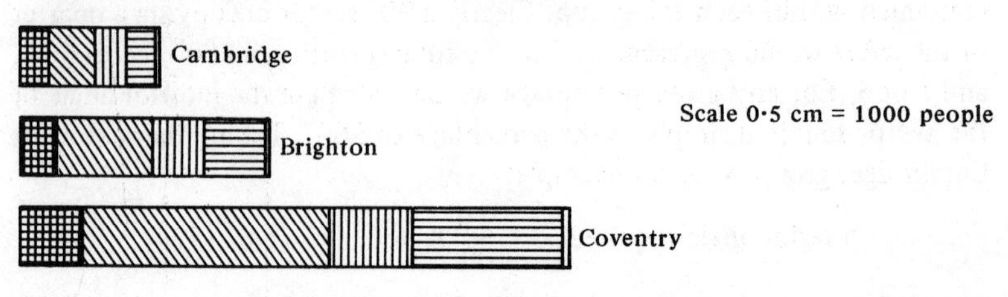

Pie graphs obviously share the advantages and disadvantages of simple proportional circles. Their chief advantage lies in their suitability for use as map symbols, relating either to points or to areas. Frequently, however, they are used as statistical diagrams in their own right, illustrating data which need not be considered within a spatial context. This is true of the example given, although if necessary the circles could be drawn on a map to show the location of the three towns. If mapping is not necessary, it is worth remembering that pie graphs are not the easiest of proportional symbols to construct. In such cases the data may often be presented as effectively, and yet more simply, by using divided proportional bars (see Fig. 8.7).

Graphs

These are versatile and commonly used statistical diagrams of relatively simple construction. They differ from symbols in that their greater bulk makes them less suitable for representing data on maps. They allow quantitative information relating to a wide range of geographical phenomena to be scaled against time, distance, or some other variable. Several different types exist, but five are of particular value to the geographer.

Fig. 8.8 *Bar graphs used to represent two different sets of data. Diagram (a) illustrates the Gross National Product per head in 1969 for selected countries. Diagram (b) shows the monthly distribution of rainfall for Cambridge. Note that values may be scaled along either axis.*

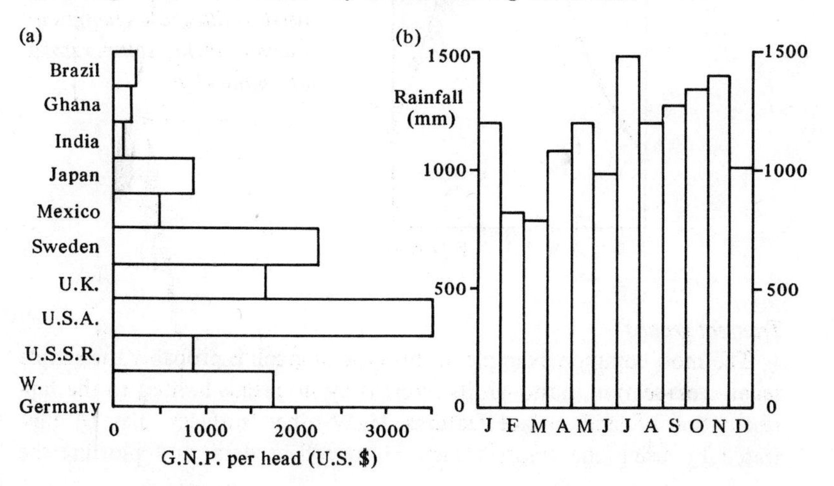

Bar graphs

These are very simple graphs which consist of a number of proportional bars of equal width and variable length. Variations in quantity are scaled along one axis only, and the bars themselves can be used to represent a wide range of variables, including places, areas, items of different types, and periods of time. Two examples of bar graphs are shown in Fig. 8.8.

Bar graphs may be used as map symbols as alternatives to pie graphs, but their bulk and clumsiness restrict their visual impact. Despite their similarity of appearance, bar graphs should not be confused with histograms (see Chapter 2), which have a quantitative scale along both axes.

Straight-line graphs

These are generally used to plot changes in a variable through time, with data available at constant intervals. An example is given in Fig. 8.9, which shows monthly variation in temperature. Points are connected by straight lines to indicate the fluctuation in value through time. With some data trends through time may be identifiable by calculating and plotting running means in the manner described in Chapter 2 (see p.20).

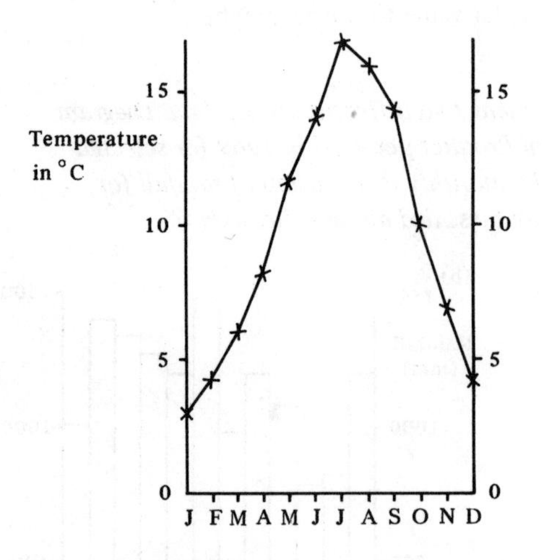

Fig. 8.9 *A straight-line graph used to illustrate changes in mean monthly temperature at Cambridge*

Transect graphs

The most common example of this type of graph is probably the simple relief cross-section, although its scope is by no means limited to the representation of land surface features. Many other 'surfaces' may be illustrated by using the transect method to collect data, and plotting the

Fig. 8.10 *An urban transect graph for part of Catford shopping centre. Information on ground floor use of premises has been omitted. For description of the methods used in collecting data of this sort see Chapallaz, D.P.* et al., Hypothesis testing in field studies, Teaching Geography *series No. 11, (Geographical Association, 1970)* **(Source:** Field work conducted by M. Callaghan, P. John, T. Snelling, and N. Swanson)

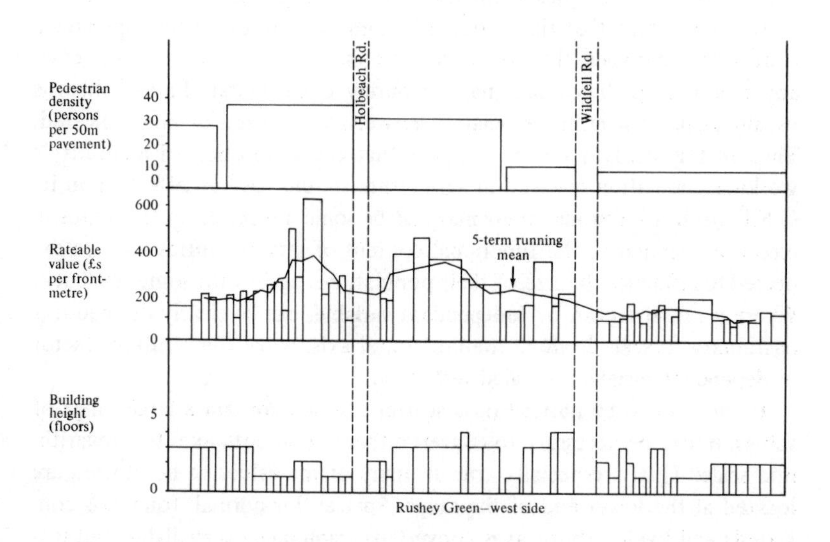

information in the form of a graph with distance scaled along the horizontal axis.

The essential feature of any cross-section or transect is that it represents variation in value or quantity along a line, which may or may not be straight. This is illustrated in Fig. 8.10, which is an 'urban transect' taken along one side of a street in the centre of Catford. Data relating to building height and rateable value were collected for each premises, along with information on ground floor use. Pedestrian densities were calculated for each length of pavement by taking moving pedestrian counts. In each case the data are plotted against the relevant section of the horizontal scale. A 5-term running mean was used to smooth out some of the variation in rateable values, standardised on a metre-frontage basis, and the results are also plotted on the transect. The graph gives some indication of the peaking of pedestrian densities, rateable values, and building heights associated with the most accessible parts of a shopping centre

Scatter diagrams

These are used to investigate the relationship between two different sets of data. They differ from the graphs already described in that variable quantities are scaled along both axes. Each item considered must have two values, one from each set of data, which serve as x and y coordinates to enable the item to be plotted as a point on the graph. This is illustrated in Fig. 8.11, which gives two examples of the use of scatter diagrams.

It is important that there should be some logical reason for expecting a relationship between the two variables concerned. If this is not the case, any relationship discovered may be purely coincidental. In fact there is usually a cause and effect association between the two sets of values plotted. Thus in Fig. 8.11(a) we may expect that the proportion of a country's working population engaged in agriculture should have an effect upon its G.N.P. per head, although there may also be some 'feedback' in the opposite direction. Similarly, the functional content of service centres may be expected to influence the size of their populations, again with some feedback. Where a causal factor, or **independant variable**, can be clearly defined it is customary to scale it along the horizontal axis, while the resultant factor or **dependent variable**, is scaled vertically.

If the data to be plotted on a scatter diagram contain a large range of values, it may be necessary to convert either one or both axes to a logarithmic scale. This is especially true if many of the values to be plotted are located at the lower end of the range. Special 'log-normal' (one axis converted) and log-log (both axes converted) graph paper is available, but it is quite possible to construct a logarithmic scale on ordinary arithmetic graph paper by plotting the logarithms of the values concerned. This is illustrated in Fig. 8.11(b), where the vertical axis is logarithmic. Although the scale is labelled normally, it is in fact the logarithm of the population of each centre that is plotted on the graph. Note that the effect of using a log-scale is to compress the upper part of the scale by progressive amounts relative to the lower part, and this makes it suitable for use with data containing a large range of values (see column headed 'difference').

Scatter diagrams allow the degree of relationship between two variables to be judged subjectively. If a definite trend exists in the distribution of points on the graph, then a relationship of some sort has been recognized. Fig. 8.11(a) illustrates an **indirect relationship**, in which one variable increases in value as the other decreases and vice versa. On the other hand, diagram (b) shows a **direct relationship**, in which both variables increase or decrease in conjunction with each other.

The closer the points on a scatter diagram are to lying on a single straight

Fig. 8.11 *Examples of scatter diagrams. Graph (a) is self-explanatory and is plotted on arithmetic axes. Graph (b) shows the relationship between population and functional content of service centres in East Anglia. It is 'log-normal', the vertical axis being scaled logarithmically*

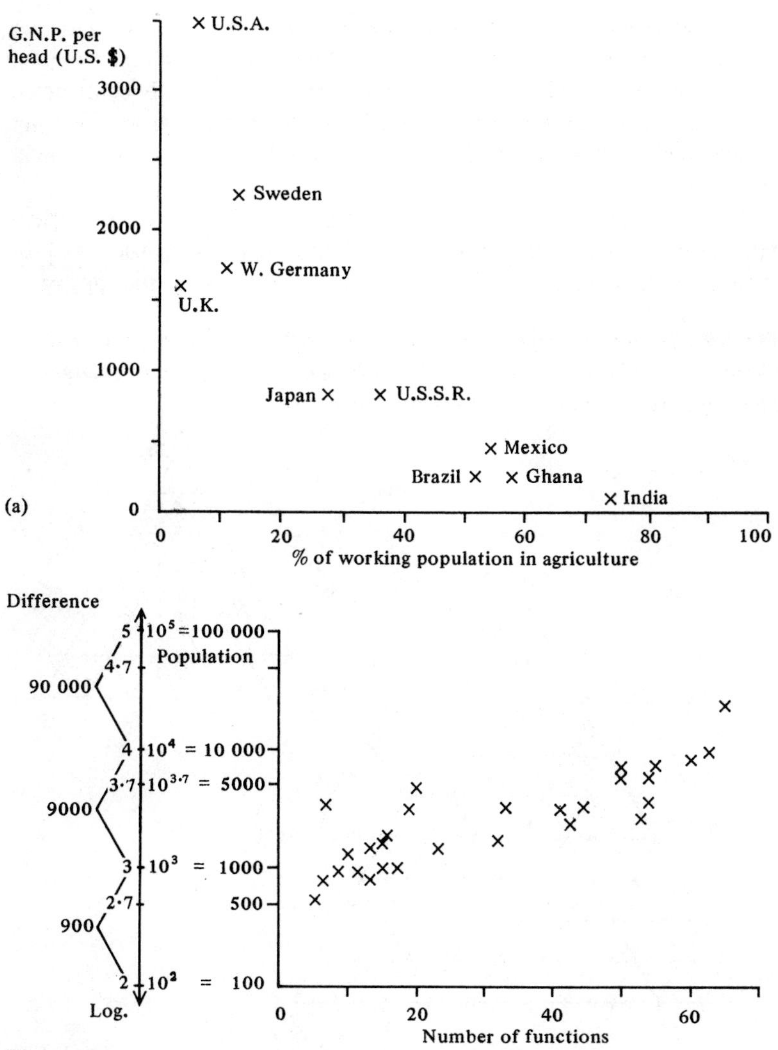

line, the greater is the degree of relationship, or **correlation**, between the two variables. It is not only possible, but also generally desirable, to make an objective assesment of the level of correlation by using a simple statistical technique, and both this and further work involving the analysis of scatter diagrams is covered in S.I.G. 4.

Triangular graphs

It is sometimes necessary to consider values of more than two variables at the same time. This presents greater difficulty with graphical representation, but the triangular graph is a convenient and fairly easily constructed means of plotting three connected variables on the same diagram. The form of the graph is illustrated in Fig. 8.12, which is based on the data given in Fig. 8.13.

The values represented by each point are scaled along three axes, orientated at 60° to each other in the form of an equilateral triangle. The position of each point is fixed in relation to the intersecting triangular grid

Fig. 8.12 *Triangular graphs used to illustrate differences in employment characteristics between regions of Great Britain. Graph (b) is an enlargement of part of graph (a). For key see Fig. 8.13*

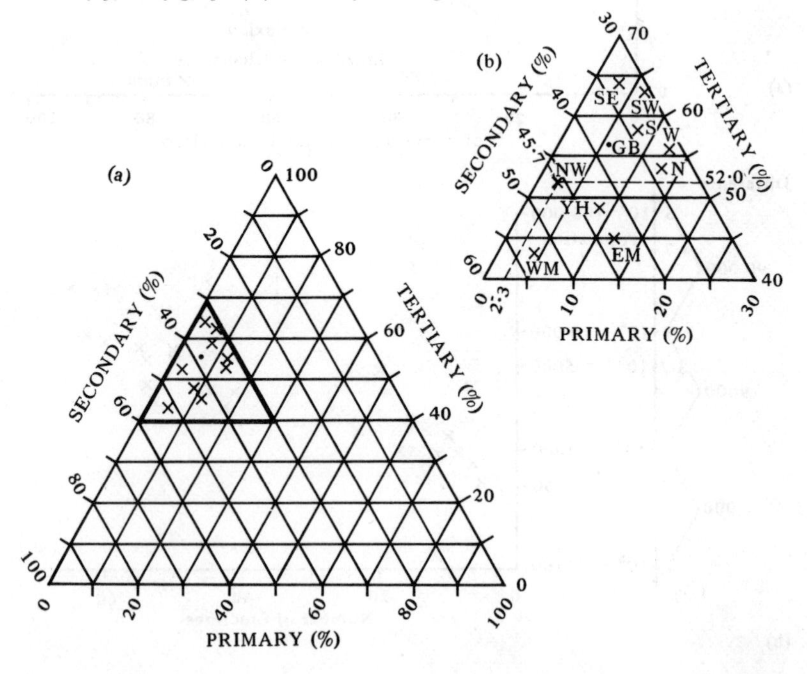

Fig. 8.13 *Employment characteristics for Great Britain by economic planning regions. The table gives the % of the total employment in each region provided by each of the three main industrial sectors. For a map of the regions used see Fig. 8.14*
(**Source**: Statistics published by Ministry of Labour)

Region	Regional employment by major categories (%)		
	Primary	Secondary	Tertiary
S.E. England (SE)	2·6	33·0	64·4
S.W. England (SW)	6·3	30·1	63·6
W. Midlands (WM)	4·2	52·6	43·2
E. Midlands (EM)	11·5	43·5	45·0
Yorkshire & Humberside (YH)	8·3	43·2	48·5
North-west (NW)	2·3	45·7	52·0
N. England (N)	12·9	33·7	53·4
Wales (W)	12·8	31·4	55·8
Scotland (S)	7·7	33·5	58·8
GREAT BRITAIN (GB)	5·5	38·0	56·5

lines marked. It is important to follow the correct grid line to the appropriate axis when locating a point, or reading off a value (see Fig. 8.12(b)).

In the example given the axes are scaled to represent employment in primary (extractive), secondary (manufacturing), and tertiary (service) industries. As Fig. 8.12(a) shows, all the points lie within a fairly small part of the graph, and to emphasize the contrast between different regions, this part has been enlarged in diagram (b). In general we can say that the emphasis on services is greatest in those regions closest to the top of the graph, while it is placed on manufacturing towards the bottom-left, and extractive industries towards the bottom-right. Obviously regions which occur close to each other on the graph have similar employment characteristics, while the greatest differences are between those regions spaced furthest apart.

The data plotted on triangular graphs must be related so that for each point it is possible to predict the third value if two are known. This is achieved by expressing the values represented by each point as percentages of their common total. Thus if we know, for instance, that 5·5% of the working population of Great Britain are employed in primary industries, and a further 38·8% in manufacturing, it follows that the remaining 56·5% must be employed in services. Any increase in one percentage must then be balanced by a corresponding decrease in one, or both, of the other two. Thus triangular graphs can only be used to represent data which can be expressed conveniently in percentage form (employment, land use, etc.).

Exercises

1. Construct a map to show the relative importance of service centres in part of East Anglia. Draw proportional circles for each of the centres included in Fig. 3.8, and use them to represent the functional index data given in Fig. 5.6 (see p.63).
2. Trace the outline shapes of the economic planning regions shown in Fig. 8.14. With the aid of the data given in Fig. 8.13, construct a map which uses divided proportional circles to illustrate (a) the total employment in each region, and (b) the distribution of employees between primary, secondary, and tertiary activities.
4. Obtain data for population density in Norwich by enumeration areas from Fig. 6.1, and for the percentage of non-manual workers from Fig. 6.11. Compare the two sets of data by plotting them on a scatter graph, using arithmetic scales on both axes. Which variable will you scale along the hori-

Fig. 8.14 *Economic planning regions of Great Britain. The total employed population of each region is given in millions of employees*

zontal axis, and why? What sort of relationship does the graph indicate? Does there appear to be a good correlation?

5. Using logarithmic scales on both axes, construct a scatter diagram to show the relationship between functional index and population for the centres listed in Fig. 5.6 (see p.63). What sort of relationship is indicated? Comment on the degree of correlation, and suggest possible explanations for any points which lie away from the general trend.

6. Using the triangular graph in Fig. 8.12 as evidence, divide the economic planning regions of Britain into groups based on their employment characteristics. Represent your grouping by using a suitable colour scheme to shade in regions of each group on a map based on Fig. 8.14.

References

1. Dickinson, G.C., *Statistical Mapping and the Presentation of Statistics* (Arnold, 1963), p. 87.

Further reading for Chapter 8

Dickinson, G.C., *Statistical Mapping and the Presentation of Statistics* (Arnold, 1963), Chapter 2, especially pp. 17–35, 38–44, and 59–63.

Monkhouse, F.J., and Wilkinson, H.R., *Maps and Diagrams* (Methuen, 1963), especially pp. 23–31, 235–54, 297–301, and 305–30.

Toyne, P., and Newby, P.T., *Techniques in Human Geography* (Macmillan, 1971), Chapter 3, especially pp. 67–84.

Bibliography

+ Cole, J.P., and King, C.A.M., *Quantitative Geography* (Wiley, 1968).
* Dickinson, G.C., *Statistical Mapping and the Presentation of Statistics* (Arnold, 1963).
* Gregory, S., *Statistical Methods and the Geographer* (Longman, 1968).
+ Haggett, P., *Locational Analysis in Human Geography* (Arnold, 1965).
+ King, L.J., 'A Quantitative Expression of the Pattern of Urban Settlements in selected areas of the United States, in Ambrose, P. (ed), *Analytical Human Geography* (Longman, 1969).
* Monkhouse, F.J., and Wilkinson, H.R., *Maps and Diagrams* (Methuen, 1963).
* Moroney, M.J., *Facts from Figures* (Penguin, 1956).
* Theakstone, W.H., and Harrison, C., *The Analysis of Geographical Data* (Heinemann, 1970).
* Toyne, P., and Newby, P.T., *Techniques in Human Geography* (Macmillan, 1971).

* Good clear outlines of subject under discussion not too advanced for sixth-form and college students, although parts may be considered demanding.

+ Rather more advanced texts, but still worth referring to and containing sections that may be read easily.

Contents of Science in Geography, books 1, 2, and 4

S.I.G. 1 Developments in geographical method

Brian P. FitzGerald

S.I.G. 2 Data collection

Richard Daugherty

S.I.G. 4 Data use and interpretation

Patrick McCullagh